Water, Technology and the Nation-State

Just as space, territory and society can be socially and politically co-constructed, so can water, and thus the construction of hydraulic infrastructures can be mobilised by politicians to consolidate their grip on power while nurturing their own vision of what the nation is or should become. This book delves into the complex and often hidden connection between water, technological advancement and the nation-state, addressing two major questions. First, the arguments deployed consider how water as a resource can be ideologically constructed, imagined and framed to create and reinforce a national identity, and secondly, how the idea of a nation-state can and is materially co-constituted out of the material infrastructure through which water is harnessed and channelled.

The book consists of 13 theoretical and empirical interdisciplinary chapters covering four continents. The case studies cover a diverse range of geographical areas and countries, including China, Cyprus, Egypt, Ethiopia, France, Nepal and Thailand, and together illustrate that the meaning and rationale behind water infrastructures goes well beyond the control and regulation of water resources, as it becomes central in the unfolding of power dynamics across time and space.

Filippo Menga is a Lecturer in Human Geography in the Department of Geography and Environmental Science at the University of Reading, UK.

Erik Swyngedouw is Professor of Geography in the Department of Geography, School of Education, Environment and Development at the University of Manchester, UK.

Earthscan Studies in Water Resource Management

The Grand Ethiopian Renaissance Dam and the Nile Basin
Implications for transboundary water cooperation
Edited by Zeray Yihdego, Alistair Rieu-Clarke and Ana Elisa Cascão

Freshwater Ecosystems in Protected Areas
Conservation and management
Edited by Max C. Finlayson, Jamie Pittock and Angela Arthington

Participation for Effective Environmental Governance
Evidence from European Water Framework Directive Implementation
Edited by Elisa Kochskämper, Edward Challies, Nicolas W. Jager and Jens Newig

China's International Transboundary Rivers
China's International Transboundary Rivers
By Lei Xie and Jia Shaofeng

Urban Water Sustainability
Constructing Infrastructure for Cities and Nature
Sarah Bell

The Biopolitics of Water
Governance, Scarcity and Populations
Sofie Hellberg

Water, Technology and the Nation-State
Edited by Filippo Menga and Erik Swyngedouw

Revitalizing Urban Waterway Communities
Streams of environmental justice
Edited by Richard Smardon, Sharon Moran and April Baptiste

For more information and to view forthcoming titles in this series, please visit the Routledge website: www.routledge.com/books/series/ECWRM/

Water, Technology and the Nation-State

Edited by Filippo Menga and Erik Swyngedouw

First published 2018
by Routledge
2 Park Square, Milton Park, Abingdon, Oxon OX14 4RN

and by Routledge
711 Third Avenue, New York, NY 10017

Routledge is an imprint of the Taylor & Francis Group, an informa business

© 2018 selection and editorial matter, Filippo Menga and Erik Swyngedouw; individual chapters, the contributors

The right of Filippo Menga and Erik Swyngedouw to be identified as the authors of the editorial material, and of the authors for their individual chapters, has been asserted in accordance with sections 77 and 78 of the Copyright, Designs and Patents Act 1988.

All rights reserved. No part of this book may be reprinted or reproduced or utilised in any form or by any electronic, mechanical, or other means, now known or hereafter invented, including photocopying and recording, or in any information storage or retrieval system, without permission in writing from the publishers.

Trademark notice: Product or corporate names may be trademarks or registered trademarks, and are used only for identification and explanation without intent to infringe.

British Library Cataloguing-in-Publication Data
A catalogue record for this book is available from the British Library

Library of Congress Cataloging-in-Publication Data
Names: Menga, Filippo, editor. | Swyngedouw, E. (Erik), editor.
Title: Water, technology and the nation-state / edited by Filippo Menga
    and Erik Swyngedouw.
Description: Abingdon, Oxon ; New York, NY : Routledge, 2018. |
    Series: Earthscan studies in water resource management | Includes
    bibliographical references and index.
Identifiers: LCCN 2017057707 | ISBN 9781138724655 (hbk) |
    ISBN 9781315192321 (ebk)
Subjects: LCSH: Water resources development—Government policy—Case
    studies. | Water resources development—International cooperation—Case
    studies. | Nation state—Case studies. | Nation building—Case studies.
Classification: LCC HD1691 .W3654 2018 | DDC 333.91—dc23
LC record available at https://lccn.loc.gov/2017057707

ISBN: 978-1-138-72465-5 (hbk)
ISBN: 978-1-315-19232-1 (ebk)

Typeset in Bembo
by Apex CoVantage, LLC

Dedicated to all those who struggle for the waters their state has taken away.

"*Water, Technology and the Nation-State* is an extraordinary and path-breaking masterpiece on the political ecologies of state-production and resistance – intellectually rich, socially urgent and politically highly revealing. The book presents a careful, critical analysis of how flows of water and power interconnect technology, nature and society. In a sophisticated way, Menga, Swyngedouw, and their impressive assemblage of authors, scrutinize and illuminate the multi-dimensional interdependence among technological trajectories, hydro-territorial configurations and nation-state building. Constituting a powerful critique of neoliberal water governance and water's de-politicizing expert-thinking, the book also offers crucial water-for-thought for building alternative hydrosocial territories."

Rutgerd Boelens, *CEDLA/University of Amsterdam,*
The Netherlands

"The world's water crisis is not only an issue of physical scarcity and declining water quality, rather it is a complex suite of social, political, and economic issues that are deeply rooted in power and the state. Menga and Swyngedouw's thought-provoking edited volume brings together a highly talented and diverse group of scholars and practitioners that explore the inter-connection between water, technology and the nation-state. This book is a must-read for anyone interested in transcending national rhetoric around water, and working towards water justice and equity."

Emma S. Norman, *Northwest Indian College, USA*

"*Water, Technology and the Nation-State* arrives at an opportune time for water-society scholars and practitioners interested in the profound ways that technological innovations have concurrently shaped waterscapes, state practices, and national identities over the past several decades. A diverse group of critical thinkers infuse cases of large-scale infrastructure development – spanning Asia, the Middle East, North America, Europe, and East Africa – with novel and incisive analyses of how water technologies are always and everywhere political and often fundamental to the exercise of state power. Combining conceptual muscle with heretofore rare case studies, this volume adds immeasurably to theories of state-nature relations and offers concrete instances of the myriad ways that dams, irrigation and hydropower have become hegemonic, and often domineering, technological interventions in human relations with water."

Christopher S. Sneddon, *Dartmouth College, USA*

Contents

List of figures	ix
Acknowledgments	x
List of contributors	xi

1 States of water — 1
FILIPPO MENGA AND ERIK SWYNGEDOUW

2 The ocean bountiful? De-salination, de-politicisation, and binational water governance on the Colorado River — 19
JOE WILLIAMS

3 Piercing the Pyrenees, connecting Catalonia to Europe: the ascendancy and dismissal of the Rhône water transfer project (1994–2016) — 34
SANTIAGO GOROSTIZA, HUG MARCH AND DAVID SAURÍ

4 Death by certainty: the Vinça dam, the French state, and the changing social relations of the irrigation of the Têt basin of the Eastern French Pyrénées — 49
JAMIE LINTON AND ETIENNE DELAY

5 Big projects, strong states? Large-scale investments in irrigation and state formation in the Beles Valley, Ethiopia — 65
EMANUELE FANTINI, TESFAYE MULUNEH AND HERMEN SMIT

6 Water nationalism in Egypt: state-building, nation-making and Nile hydropolitics — 81
RAMY HANNA AND JEREMY ALLOUCHE

7 Troubled waters of hegemony: consent and contestation in Turkey's hydropower landscapes — 96
BENGI AKBULUT, FIKRET ADAMAN AND MURAT ARSEL

viii *Contents*

8 **An island of dams: ethnic conflict and the contradictions of statehood in Cyprus** 115
PANAYIOTA PYLA AND PETROS PHOKAIDES

9 **Counter-infrastructure as resistance in the hydrosocial territory of the occupied Golan Heights** 131
MUNA DAJANI AND MICHAEL MASON

10 **Development initiatives and transboundary water politics in the Talas waterscape (Kyrgyzstan-Kazakhstan): towards the Conflicting Borderlands Hydrosocial Cycle** 147
ANDREA ZINZANI

11 **Speculation and seismicity: reconfiguring the hydropower future in post-earthquake Nepal** 167
AUSTIN LORD

12 **Irrigational illusions, national delusions and idealised constructions of water, agriculture and society in Southeast Asia: the case of Thailand** 189
DAVID J.H. BLAKE

13 **Building a dam for China in the Three Gorges region, 1919–1971** 207
COVELL F. MEYSKENS

Index 223

Figures

4.1	Map of the Department of Eastern Pyrénées, southeastern France	52
4.2	Two patterns of social relations structured by irrigation	60
5.1	Location of the Beles Sugar Development Project and the main zones of labour migration towards it	68
7.1	Çoruh Basin Project	106
8.1	Map of various types of dams constructed in Cyprus	121
8.2	United Nations military personnel assisting villagers transfer water and goods with a donkey across division lines in Cyprus	122
9.1	Shallow pools and rainwater harvesting tanks	141
9.2	The upscaling of counter-infrastructure through the construction of community-funded pipelines and pumps	142
10.1	GIS elaboration of a satellite image representing the upstream part of the Talas waterscape and borderlands, focus of ethnographic research	153
10.2	GIS elaboration of a satellite image representing the Talas waterscape borderlands, the Talas River	159
12.1	Representation of a range of hydraulic infrastructure development options proposed, planned and/or implemented at different time periods in the Northeast	194

Acknowledgments

This book is the result of a collective effort that originated at the "Interdisciplinary Workshop on Water, Technology and the Nation-State", which took place at the University of Manchester in October 2016. This two-day workshop brought together over 50 scholars and students from around the world that discussed and tried to disentangle the complex and often hidden connection between water, technological advancement and the nation-state. We are therefore grateful to all those who participated and contributed with their research, ideas and comments. The workshop would not have been possible without generous support from the European Union's Horizon 2020 research and innovation programme under the Marie Sklodowska-Curie actions (grant number 654861), for which we also are grateful.

A big thank you goes to the 22 authors who contributed to this volume, sharing with us their research and rich empirical material and thus making this book come to life. We hope that this volume will stimulate future collaborations between some of its authors, while also consolidating existing links and transcontinental (and interdisciplinary) exchanges.

We also thank the editorial team at Routledge Earthscan, and in particular Tim Hardwick for inviting us to put this volume together, and Amy Johnston, who assisted us during the whole editorial process. Filippo Menga also thanks Elsevier for allowing him to reprint parts of his *Political Geography* article "*Hydropolis*: Reinterpreting the *polis* in water politics" (Volume 60, September 2017) and the Water Alternatives Association for allowing him to reprint parts of his *Water Alternatives* article "Domestic and international dimensions of transboundary water politics" (Volume 9, Issue 3, 2016).

Contributors

Fikret Adaman (BA and MA in economics at Boğaziçi University and PhD in economics at University of Manchester) is currently professor of economics at Boğaziçi University, Istanbul, Turkey. His research interests include alternative economies, ecological economics, history of economic thought and political economy. His publications have appeared in, among others, *Antipode, Cambridge Journal of Economics, Conservation Letters, Development and Change, Ecological Economics, Energy Policy, Environmental Politics, New Left Review* and *Voluntas*. His co-edited book *Neoliberal Turkey and its Discontents: Economic Policy and the Environment Under Erdoğan* has recently been published (IB. Taurus). He has been acting as expert to the European Commission for Turkey on social exclusion since 2009, as director of the Turkey office of UN-Sustainable Development Network Solutions since 2016, and as a member of UNESCO Sustainable Development Unit, Turkey, since 2017. He served as chair to the Economics Department and as advisor to the rector of Boğaziçi University.

Bengi Akbulut (BA Economics, Bogazici University; PhD Economics University of Massachusetts at Amherst) is an Assistant Professor at the Department of Geography, Planning and Environment, Concordia University. Her work lies within the fields of political economy, ecological economics, development studies and feminist economics. Her joint and independent work has appeared in the *Cambridge Journal of Economics, Development and Change* and the *Journal of Peasant Studies* among others. Most recently she has co-edited a volume on the political ecology of neoliberal Turkey, *Neoliberal Turkey and its Discontents: Economic Policy and the Environment Under Erdoğan* (I.B. Tauris).

Jeremy Allouche is currently a Research Fellow and cluster leader of the Resource Politics cluster at the Institute of Development Studies, University of Sussex, and a member of the ESRC-funded STEPS (Social, Technological and Environmental Pathways to Sustainability) Centre, the Sussex Center for Conflict and Security Research and the Sussex African Center. He previously worked at the University of Oxford, the Massachusetts Institute of Technology, ETH Lausanne, at the Swiss Graduate Institute of Public Administration and the Graduate Institute of International and Development Studies. His xii *Contributors*

fields of interests are the intersection between global environmental change, development and security.

Murat Arsel is Associate Professor of Environment and Development at the International Institute of Social Studies (ISS) of Erasmus University Rotterdam. His research concerns the political economy of state-led development and the relationship between nature and capitalism. He is currently working on the class dynamics of climate change politics and writing a book on the Yasuni-ITT initiative of Ecuador to be published by Oxford University Press.

David J.H. Blake is an independent scholar based in the United Kingdom with a professional background in environmental management and rural development in Thailand and Laos. He holds a PhD from the School of International Development, University of East Anglia, with a thesis examining the socio-political drivers of irrigation development in Thailand's Northeast. He was the Luce Visiting Scholar in Environmental and Urban Studies at Trinity College, Hartford, USA in the autumn term of 2016. Currently he is researching a number of themes around water resources governance in the Mekong Region.

Muna Dajani is a PhD candidate in the Department of Geography and Environment at the London School of Economics and Political Science (LSE). Her research examines decades of Israeli water and land policies and their impact on the livelihoods and lived experiences of Palestinian and Syrian agricultural communities and how these communities contest these policies. Her experience with research projects includes working on the *Hydro-political Baseline of the Upper Jordan* (UEA, 2011), *Transboundary Climate Security: Climate Vulnerability and Rural Livelihoods in the Jordan River Basin* (LSE and Birzeit University, 2014) and the *Hydro-political Baseline of the Yarmouk River* (UEA, 2017). She has co-authored articles in journals such as *Water Alternatives* and *International Journal of Housing Policy.*

Etienne Delay is a social geographer with a postdoctoral fellowship position at CNRS (French National Center for Scientific Research). He holds MSc and PhD degrees in geography from Limoges University. To tackle the scientific challenges proposed by landscape dynamics and cooperation processes, he has developed a research methodology based on fieldwork and companion modelling combined with agent-based modelling formalisation. He applied these methods to various contexts: steep slope vineyards landscapes (2011), water resource management cooperatives (2015) and vegetation cover in the Sahelian great green wall (2017). His main research interests concern socio-spatial interactions and agent-based simulations with a focus on co-operation emergence.

Emanuele Fantini is lecturer and researcher at IHE Delft Institute for Water Education, in the Netherlands. He holds a PhD in Political Sciences (University of Turin) and a European Master's Degree in Human Rights and

Democratisation (University of Padua). His research interests include: water governance and conflicts, social movements, religion in public spaces, media studies and visual research methods, with geographic focus on Ethiopia, the Nile basin and Italy. Emanuele is also associate researcher at the Programme in Comparative Media Law and Policy—University of Oxford and coordinator of the project "Open Water Diplomacy. Media, science and transboundary cooperation in the Nile basin", studying the role of media and scientific communication in transboundary water interactions (www.nilewaterlab.org).

Santiago Gorostiza is a postdoctoral researcher at the Institute of Environmental Science and Technology at the Universitat Autònoma de Barcelona (ICTA-UAB). He is an environmental historian working on modern and early modern Spanish history, with training as an historian, environmental scientist and political ecologist. He carried out his PhD at the Centro de Estudos Sociais of Universidade de Coimbra (Portugal) as a Marie Curie ITN fellow of the European Network of Political Ecology (ENTITLE). His doctoral research examined how the Spanish Civil War (1936-1939) and the Francoist victory and state-building efforts transformed the country's socioecological relations and landscapes, both materially and symbolically.

Ramy Hanna is a doctoral researcher at the Institute of Development Studies (IDS), University of Sussex. His PhD research encompasses resource politics, water security and transboundary water governance. His academic background is economics and international relations, and he earned his Master's degree in Environmental Studies from York University in Toronto. His research and professional experience include working with the Escuola Superior de Agricultura (ESALQ) at the University of São Paolo in Brasil, FAO (Food and Agricultural Organization of the UN) Water Scarcity Initiative, IFAD (International Fund for Agricultural Development), UN Economic Commission on Africa and WBG (World Bank Group). As an environment specialist he conducted the Nile Basin Discourse evaluation and advised the Eastern Nile Watershed Management Project.

Jamie Linton currently directs a research programme in river management at the University of Limoges, France. His main research interest is in the relations between water and people from an historical and geographical perspective. Author of *What is Water? The History of a Modern Abstraction* (UBC Press, 2010) among other publications, he is currently involved in personal and collaborative research projects focusing on various aspects of the hydrosocial cycle, public participation in river management, the memory of floods, the history of socio-fluvial relations in France, and the political implications of rethinking the materiality of water.

Austin Lord is a PhD student in the Department of Anthropology at Cornell University, USA, where his dissertation research focuses on questions related to disaster and aftermath, infrastructure development, and the lived experience of uncertainty and environmental change in Nepal. His research on water, energy and disaster in Nepal has been published in a variety of academic journals, including *Economic Anthropology*, *Eurasian Geography and Economics*, *Modern Asian Studies Himalaya: The Journal of the Association for Nepal and Himalayan Studies*, and *Limn*. Austin also holds a Master's degree in Environmental Science from Yale University, and a collection of his visual ethnographic work can be found at www.austinlord.com.

Hug March is an Associate Professor in the Faculty of Economics and Business at Universitat Oberta de Catalunya (UOC) (Spain) and researcher at TURBA Lab (IN3, UOC). His research focuses on the urban political ecology of the hydrosocial cycle and the political ecology of urban transformation and alternative socio-ecological transitions. He has extensively published his research in international journals such as *Annals of the Association of American Geographers*, *Antipode*, *Geoforum*, *IJURR*, *EURS*, *Habitat International*, *Political Geography*, *Journal of Cleaner Production* and *Journal of Hydrology*, among others.

Michael Mason is an Associate Professor in the Department of Geography and Environment at the London School of Economics and Political Science. Michael has research interests in global environmental governance – notably issues of accountability, transparency and security – and environmental politics in post-conflict societies. Alongside articles in a wide range of academic journals, he is the author of *Environmental Democracy* (1999) and *The New Accountability: Environmental Responsibility across Borders* (2005). He is also co-editor of (with Amit Mor) *Renewable Energy in the Middle East* (2009) and (with Aarti Gupta) *Transparency in Global Environmental Governance* (2014).

Filippo Menga is a Lecturer in Human Geography at the University of Reading, and his research mainly explores the politics of transboundary waters, global water governance and the interplay between nationalism and water conflicts. He has held research and teaching positions at the University of Manchester, Oxford Brookes University, Tallinn University, the University of St Andrews, King's College London and the University of Cagliari. He has published articles in a wide range of academic journals, including *Political Geography*, *Geoforum*, *Nationalities Papers* and *Water Alternatives*.

Covell F. Meyskens is an Assistant Professor at the Naval Postgraduate School. A graduate of the University of Chicago, his research focuses on war and industrial development in modern China. He is currently finishing up revisions on his first book, *The Third Front: Maoist China's Developmental Garrison State*, and he has recently started working on his second book on the history of the Three Gorges Dam. Other ongoing projects include changing notions of national security in modern China and an urban history of the Third Front town of Panzhihua.

Tesfaye Muluneh Befekadu is a Lecturer and Researcher in Geography and Environmental Studies at Wollega University in Ethiopia. He holds a BA in Geography and Environmental Studies from Bahir Dar University in Ethiopia and a MA in Geography and Environmental Studies from Addis Ababa University specializing in Regional and Urban Development Planning. His research interests include environmental and human issues at regional and urban scales, along with the utilisation of GIS (geographic information systems) and remote sensing technologies to deal with both environmental and social development planning.

Petros Phokaides is an architect and PhD candidate in History-Theory of Architecture, at the National Technical University of Athens, Greece, and a Researcher at Mesarch Lab, University of Cyprus. He is currently completing his dissertation on transnational architecture and planning, focusing on the work of Doxiadis Associates in 1960–1970s Africa. He has performed extensive archival research in Greece, Cyprus and the United Kingdom and has authored several articles published in the *Journal of Architecture*, *Docomomo Journal*, *MIT Thresholds* and *MONU Magazine*. His latest project is a co-edited volume on the role of memory and museums in contested spaces published in 2016.

Panayiota Pyla is an architectural historian and theorist and Associate Professor at the Department of Architecture, University of Cyprus. Previously, she served on the faculty of the University of Illinois at Urbana-Champaign and was a postdoctoral fellow at the Harvard Design School. Pyla holds a PhD from the Massachusetts Institute of Technology, and her work on Constantinos Doxiadis, development, sustainability, architectural modernism and historiography has been published in the *Journal of Architectural Education,* the *Journal of Architecture* and *Grey Room*, among others. She also edited *Landscapes of Development: The Impact of Modernization Discourses on the Physical Environment of the Eastern Mediterranean*, Cambridge: Harvard University Aga Khan Program, 2013.

David Saurí is Professor of Human Geography at the Universitat Autonoma de Barcelona where he teaches in the degrees of Geography, Environmental Sciences and East Asia Studies. He is primarily interested in environmental hazards, water management, tourism and climate change, and nature-society theories. He is a recipient of the Environmental Award of the Institute of Catalan Studies.

Hermen Smit is Lecturer in the Chair Group of Water Governance at the IHE Delft Institute for Water Education. His education and research examine how river morphologies (forms and infrastructures) and water users change in tandem through projects of river basin development. His research focuses on projects in the Nile basin countries in particular. With Emanuele Fantini and others he works in the nilewaterlab.org.

Erik Swyngedouw is Professor of Geography at the University of Manchester in its School of Environment and Development. He received his PhD entitled "The production of new spaces of production" under the supervision of

David Harvey at Johns Hopkins University (1991). From 1988 until 2006 he taught at the University of Oxford and was a Fellow of St. Peter's College. He moved to the University of Manchester in 2006. His research interests include political economy, political ecology and urban theory and culture. He aims at bringing politically explicit yet theoretically and empirically grounded research that contributes to the practice of constructing a more genuinely humanising geography. He is an elected member of the Academia Europaea.

Joe Williams is Assistant Professor of Human Geography at Durham University. His interests broadly encompass political ecology, development, science and technology, urban studies and political economy. Joe engages with contemporary theoretical debates in materialist social philosophy around the politics of relationality, social power, environmental transformation and posthuman thought through his research on the politics of water governance and the water-energy nexus.

Andrea Zinzani is a Research Associate at the Global Development Institute, University of Manchester. Joint PhD in Human Geography at the University of Fribourg and at the University of Verona, he worked as a Research Fellow at the University of Fribourg, at the CNRS – UMR 7528 Mondes Iranien et Indien – and at the Leibniz Center for Tropical Marine Ecology in Bremen. His research interests range from political ecology to hydropolitics and development studies, with a regional focus on Central and Southeast Asia. He published his monograph *The Logics of Water Policies in Central Asia: the IWRM implementation in Uzbekistan and Kazakhstan* (LIT Verlag, 2015), together with articles in international journals such as *Geoforum, International Journal of Water Resources Development, Cahiers d'Asie Centrale* and *Environmental Earth Sciences*.

1 States of water

Filippo Menga and Erik Swyngedouw

We live in times marked by increasing concern over an imminent water crisis. Any query of 'water wars' on an online search engine will generate thousands of hits even if the search is limited to only the most recent year. Journalists and policy analysts quarrel over whether the first water war in history will be fought, for instance, between China and India over the waters of the Brahmaputra River; between Arizona, California and Nevada for the water allocation of Lake Mead; or between Mexico and the United States over the desiccating Colorado River. The underlying assumption in many of these accounts is that sooner or later humans will have to engage in a war against droughts, as if the latter were driven by some sort of divine or natural process over which humans have no control. In addition to nurturing the view of an imminent ecological catastrophe, this representation of the dualist vision between humans versus nature overlooks the inherent power dynamics and the interconnected processes through which nature is socially and materially produced, transformed, and contested (Heynen et al., 2006). The axiom that our planet (and its human inhabitants) needs to be saved from a looming crisis is based on the notion that an originally pristine and stable nature, but now disturbed, needs to be restored to its original state (Castree, 2014), which needs to be preserved if ecological harmony is to be maintained. Although such benign original state never existed, this view also disavows that the predicted future crisis is already here for many people and places; they already live in the midst of a crisis.

Global inequality has taken extreme forms. Hundreds of millions of people are stuck in poverty while only eight individuals own the same wealth as half of the world's population (Oxfam, 2017). The world's poorest are the most vulnerable to climate change and environmental calamities caused by extreme weather events (IPCC, 2012), as demonstrated by the consequences of Super Typhoon Yolanda in the Philippines or Hurricane Katrina in New Orleans (see for instance Smith, 2006). Disadvantaged populations globally carry the burden of environmental degradation, and the water sector is no exception. The 2015 WHO report on "Progress on sanitation and drinking water" (WHO, 2015) paints a stark and disturbing picture: globally, 2.5 billion people lack improved sanitation and 1.1 billion people practice open defecation. By 2025, 1.8 billion people will be living in conditions of absolute water scarcity, as their annual water supplies will be below 500 cubic metres per person per year (to put this in context, in 2014 water withdrawals per capita in Estonia amounted to 1,231.7 cubic metres, in Greece to 869.3 cubic metres, and in Mexico to 709.4 cubic metres; OECD, no date). The two latest UN World Water Development Reports (WWAP, 2017, 2016) warned that the global water crisis is caused by poor governance rather than by resource availability, stressing the need for new, inclusive technical solutions. Yet, this is done through the adoption of a depoliticising language which nurtures and advocates a techno-managerial framing that advances economic and technological solutions, rather than politically challenging the way in which we manage and consume the planet's natural resources (Rogers et al., 2016; Williams et al., 2014).

Indeed, water is not just a natural resource and a physical agent, but it is also deeply embedded in social, political, and economic processes (Mollinga, 2008; Swyngedouw, 2006). While the former view has championed techno-cratic approaches to water management based on narratives that are primarily informed by the natural sciences (Sharp, 2017), the latter are illustrated by deepening processes of appropriation of water resources by powerful actors and the parallel dispossession of weaker or marginalized social groups (Mehta et al., 2012). Appropriating water assigns power to those who control it (Norman et al., 2015), and hydraulic infrastructures can consequently be used to wield power and to enact hegemonic and counter-hegemonic strategies. Due to its unique nature, high-level bureaucrats and technocrats tend to perceive and portray water as a national asset constituting an integral part of 'the homeland'. Just like space, territory, and society can be socially and politically co-constructed, so is water, and as a result the construction of large hydraulic infrastructures can be mobilised by politicians to consolidate their grip on power while nurturing their own vision of what the nation is or should become. This book will delve into the complex and often hidden connection between water, technological advancement, and the nation-state, addressing two major questions. First, the arguments deployed in this collection consider how water as a resource can be ideologically constructed, imagined and framed to create and reinforce a national identity, and second, how the idea of a nation-state can and is materially co-constituted out of the material infrastructure through which water is harnessed and channelled.

These questions will be addressed though a range of theoretical and empirical interdisciplinary contributions covering four continents. As the case studies will illustrate, the meaning and rationale behind water infrastructures goes well beyond the control and regulation of water resources, as it becomes central in the unfolding of power dynamics across time and space. Before providing an overview of the content of each of the chapters, this introduction delineates the main theoretical and conceptual approaches underpinning our understanding of how water resources are enmeshed with multiscalar processes related to technology and the nation-state.

Water beyond H_2O

Water is a chemical substance whose molecule is formed by two hydrogen and one oxygen atoms (H_2O). Water tends to be liquid, but can also be found in solid (ice) and gaseous (vapour) forms nearly everywhere in and on our planet. Water is thus fluid, transient, at times intangible and a ubiquitous prerequisite for life for which there is no substitute. It is perhaps for these attributes that water is arguably the most intersectional and interdisciplinary among all natural resources. Bridges are built to go over it, but water also bridges the gap between the natural and social sciences. In addition to specific water-related disciplines such as hydrology and hydraulic engineering, many other (and in certain cases less obvious) disciplines have taken an interest in water, including anthropology, science and technology studies, economics, theology, political science, international relations, law, archaeology, and geography. All of these have advanced our understanding of how the meaning and impact of water stretches well beyond its biophysical materiality. And geography, above all, has teased out the intangible and intricate web of relationships among people, places, and ideas that turns out to be central in the appropriation, dispossession, and distribution of water resources.

Karl August Wittfogel's book *Oriental despotism: A comparative study of total power* (1957) has been seminal in informing debates on the relationship between large-scale irrigation systems in arid and semi-arid regions and the consolidation of a centralised despotic political authority. While Wittfogel's environmental determinism garnered considerable criticism (Obertreis et al., 2016), the book can be praised for being theoretically insightful and for serving as a foundation that scholars have been working with and transforming his ideas during the last six decades. This has helped to disentangle the link between state formation and 'hydrosocial territories', a notion that Boelens et al. (2016: 1) deploy to define "spaces that are (re-)created through the interactions amongst human practices, water flows, hydraulic technologies, biophysical elements, socio-economic structures and cultural-political institutions". Such approach resonates with the view that nature and society are intimately interdependent and cannot be separated from one another (see Castree, 2008; and Perreault, 2014 for a critical review of the existing literature). This solidifies the view that socio-natural configurations and socio-spatial scales can be constructed and reconstructed through ideas, beliefs, and assumptions that are based on discursive, ideological, cultural, scientific, and material practices (Menga, 2017; Swyngedouw, 2010b). In relation to this, scholars have been increasingly using the concept of 'waterscape' to discuss the interactions between water, power, and socio-political dynamics at different geographical scales (Zinzani and Menga, 2017; Budds and Hinojosa, 2012; Loftus, 2009; Swyngedouw, 1997).

From the above viewpoint, a significant scalar tension comes to the fore when we analyse the state. On the one hand, the state as a political creation and an administrative body is inscribed in a relatively fixed territory whose external and internal boundaries are (in most cases) well defined, and so is its institutional and managerial configuration, or at least this tends to be the belief held by those who govern. The state is thus responsible for the control and use of its resources, including its freshwater. On the other hand, this rather traditional idea is being continuously challenged by alternative visions of the state, which emerges as one of the many highly fluid spatial scales around which intricate social interactions unfold. The nominal representation of the state as a single, monolithic object superficially conceals the intricate network of power relations and informal arrangements that underpin its functioning and everyday operation. Research has indeed shown that informal arrangements often coexist with the formal norms and rules set by the state, and water planners operate at the intersection of the two (Innes et al., 2007). As Ahlers et al. (2014) discussed in their introduction of a special issue of the journal *Water Alternatives* on informality in the urban waterscape, both formality and informality are fluid concepts, and the distribution and control of water resources depend on a dynamic set of social and material interactions which are mediated by technological development and take place at multiple scales. Formal governments can rely on informal arrangements whereby the use of land and its resources is allocated to new users and owners, based on an arbitrary unmapping of territory, thus denoting the "territorial impossibility of governance" (Roy, 2009: 81).

Highly emotionally charged and symbolically powerful hydraulic infrastructures arguably occupy a topological space of exception in which the state is seen to operate as a unified and coherent constellation and through which it both demonstrates, performs, and consolidates its power (Agamben, 2005). If we accept that the state is a dynamic geographical construction revolving around choreographies of power that need to be constantly actualised to nurture and sustain its existence, we can also appreciate how ruling politicians can and do transform the physical space occupied by a hydraulic infrastructure into a political one "whose spatial extent cannot be demarcated in any way other than by that defined by their space of appearance" (Menga, 2017: 102). Hydraulic infrastructures thus emerge as one of the ways in which the state actualises power over its territory, and therefore, also as one of the ways in which this power can be contested. Conceptual difficulties in defining the state can be addressed by emphasising how life is permeated by social relations of stateness in ordinary, if not prosaic, ways (Painter, 2006). It seems analytically relevant to consider the state as an assemblage (Dittmer, 2014), rather than a thing in itself; as an heterogeneous grouping of actors and forces operating at the social, political, and economic levels, eventually leading to the construction of a national network of interests within specific historical contexts (Swyngedouw, 2015). We shall develop this view through the ways in which this is reflected in the transformation and manipulation of the waterscape. Waterscapes (and territories) not only coexist with the state at several scales, but the two are interdependent, and produce and reproduce each another in a mutually constitutive manner.

The state is not the only conceptually fuzzy term that we shall encounter in this volume. We are also concerned with the nation. Both often fuse together and are occasionally used interchangeably to define countries as political and administrative entities. While it is not our intention to engage in an extensive review of the literature on state- and nation-building, and the nation-state, a few crucial observations need to be made explicit. On the one hand, state-building refers to the processes through which the state, as a clearly defined administrative entity, establishes the range of social institutions necessary for its functioning, including a constitution (with a few exceptions such as the U.K.) and legal framework, a government, and other state institutions such as the army and the judiciary system, which are ultimately essential in asserting its monopoly over violence and maintaining territorial sovereignty (Weber, 1978). Lately, the term state-building has been increasingly used to refer to the efforts made by the United Nations (UN) and other international organisations to re-build the abovementioned institutions in post-conflict and transitional countries (Chesterman, 2004). On the other hand, nation-building is a widely debated notion that, in general terms, refers to the set of both top-down and bottom-up initiatives, policies, ideas, and imaginaries aimed at creating a common national identity and a sense of patriotism and loyalty toward the state (Menga, 2015). The implication with this term is that even if the nation is an immaterial entity, a social or cultural construct, which can be interpreted as an imagined politi-cal community (Anderson, 2006), this does not mean that a nation cannot be constructed or built. Furthermore, this also implies that both states and nations have to be understood as processes that can be fixed in particular moments in time, rather than as pre-existing entities (Kuus and Agnew, 2008).

If we are to link the two, and they are indeed inextricably mutually constituted, it seems appropriate to follow Giddens's (1981: 182) terminology and use the term nation-state to emphasise their enmeshment and to delineate the post-war period marked by the successes of capitalism and its "eventuating in the world-wide triumph of the nation-state as a focus of political and military organisation" with its associated monopoly on violence. In this evolutionary-transformative process, Giddens (1994) also suggested that globalisation, modernity, and the ecological crisis brought with them a new scenario marked by the waning of traditional political ideologies, whereby the influence of the nation-state is being eroded by the new global agenda and the emergence of transnational actors.[1] This resonates with studies illustrating the consolidation of a 'post-political condition', one in which the people's capacity to deliberate and act 'politically' is being foreclosed by assumptions about the inevitability of liberal democracy, rule by expert knowledge, and imposed and unequally constituted cosmopoli-tanism (Swyngedouw, 2010a; Mouffe, 2005; Žižek, 2000). And yet, this same arrangement is paradoxically being challenged, in the name of democracy, by the Occupy movement and other popular protests (Wilson and Swyngedouw, 2014), while also being increasingly questioned by a new rise of nationalist and populist movements, both in the Global North and in the Global South.

As this volume will illustrate, nationalism can also be channelled into hydrau-lic infrastructures, reconfiguring the hydrosocial cycle and power dynamics in society. In their contribution to this volume, Linton and Delay (2018) argue that "the nation-state is less a thing in itself than a network of heterogeneous actors that might be considered a relational accomplishment, something that is continually affecting socionature and changing in relation to changes in socionature". There is indeed a myriad relations connecting technology with nature and society (Obertreis et al., 2016), and technological innovation has been

6 *Filippo Menga and Erik Swyngedouw*

studied as the benchmark by which nation-states enact claims of modernisation and progress. The ideologisation of technology can also be perceived as a means through which ruling elites overcomplicate technology and practical questions in such a way that the population is depoliticised and stripped of its participatory democratic rights (Habermas, 1970). As this book shall demonstrate, such a relational accomplishment is often closely interrelated with an idea of technology and progress falling under the premises of a coveted hydrological modernisation (Kaika, 2005).

When it comes to the issue of large dams, for example, the World Commission on Dams (WCD)[2] noted that "[f]rom the 1930s to the 1970s, the construction of large dams became – in the eyes of many – synonymous with development and economic progress. Viewed as symbols of modernisation and humanity's ability to harness nature, dam construction accelerated dramatically." (WCD, 2000: xxix). In this regard, Worster (1984) mobilised the Hoover Dam in the United States as an emblematic example to note that large dams have been built following the illusion that men can dominate nature. Drawing on Horkheimer's (1974) *Eclipse of Reason*, Worster argued that dominating nature also implies dominating men, since a few powerful individuals manage to concentrate significant social, economic, and political power through the construction of a dam. Work in political ecology has further underlined the complex pattern through which water (and in this case modern water supply systems) is part of a complex network of economic powers and interest groups that erode the centralised power of the government (Gandy, 2003). Likewise, work in environmental history has illustrated how a river can acquire ontological relevance in relation to the broader processes through which technology can transform nature, and with it, society (White, 1995).

Politics of water in space and time

The above discussion foregrounds a series of interrelated questions that deserve to be addressed, and that provide the analytical outline for the chapters forming this book. With both water and the nation-state being fluid and transient entities, we need to understand what spatial scales are produced and contested by the interaction between humans and water. Significant research has been carried out in political geography (among others, Herod and Wright, 2008; Flint and Taylor, 2007; Newman and Paasi, 1998; Cox, 1998; Delaney and Leitner, 1997) and political ecology (Neumann, 2009; Brown and Purcell, 2005; Swyngedouw and Heynen, 2003; Swyngedouw, 1997; Blaikie and Brookfield, 1987), to understand the multiscalar interplay between transient natural resources and the political constructs – such as countries, institutions, and borders – that have to manage them. In this regard, seminal research by John Agnew (2010, 1994) warned about the risk of falling into the territorial trap that relates to three geographical assumptions which laid the theoretical foundation for the three mainstream ontologies in international relations theory (the realist, the neo-realist, and the liberal). The first assumption is that states are fixed units of sovereign space. The second is that the domestic is separated from the foreign, while the third is the assumption that the state existed prior to, and as a container of, society (Agnew, 1994).

For the purposes of the present volume, therefore, it is necessary to clarify further that the nation-state is understood as a heterogeneous assemblage of social groupings, political actors, and economic forces, as this can facilitate our understanding of the intricate and intangible web of relationships that play a crucial role in the appropriation, dispossession, and distribution of water resources. Space, territory, and society are materially, socially, and politically constructed, and various scales of analysis need to be carefully considered to understand the politics of water (Norman et al., 2015; Norman et al., 2012; Harris and Alatout, 2010; Furlong, 2006; Sneddon and Fox, 2006; Harris, 2005). The nation-state scale, for instance, cannot be studied without the interstate (or international) scale and the basin-regional scale. As Harris puts it, "each of these functional scales can be understood in isolation, but can also be understood as being linked to processes, actors, and systems across all other scales of analysis" (Harris, 2005: 267). Political constructions of scale play a role in the management and sharing of water resources, and different discourses and strategies can be constructed and adopted at different scales. As research in critical water geography has illustrated, rivers are discursively constructed as complex sociotechnical processes (Akhter, 2015) and also as unique spatial entities where geopolitical objectives unfold and are imposed on citizens, usually through top-down means (Sneddon and Fox, 2015). And yet, while this seems to imply that ruling elites are able to predate political space to foreclose political encounter in the 'high politics' of water (Menga, 2017), Norman and Bakker (2009) have convincingly demonstrated that a shift in scale downward to the subnational level does not necessarily lead to greater empowerment for local actors, and we therefore have to avoid yet another territorial trap, the local one. Issues related to rescaling transboundary water governance have thus emerged (Norman, 2014), along with increased attention to the policy challenges stemming from global water governance (Gupta et al., 2013).

As the chapters in this book will illustrate, scalar politics of water governance are being continuously challenged and reconfigured around water infrastructures, and this happens in highly conflictual settings where struggles for social power unfold and evolve over time; from imagined to existing canals, small- and large-scale irrigation projects, diverted rivers and oceans commodified through modern desalination technologies, hydrosocial territories are being constituted and reconstituted and water as a resource becomes part of processes of hydrosocial transformation. The chapters (see in particular chapters 3, 4, 5, and 9) will also underline how hydraulic infrastructures can play a significant role in challenging, but also in consolidating, the centralised authority of the state and its territorialisation. While all the empirical evidence presented refers to the contemporary era, the chapters touch upon different times and focus on the use and availability of both traditional (irrigation and river diversion) and relatively modern (large-scale hydropower and desalination plants) technologies to manage and exploit water resources. Under a historical materialist perspective, this sheds light on the predatory nature of capitalism and on the processes and dynamics through which hegemonic economic and political elites seize hydraulic technologies to maintain and consolidate their grip on power over time. As Harvey observed, "[t]he capitalist operates in continuous space and time, whereas the politician operates in a territorialized space and, at least in democracies, in a temporality dictated by an electoral cycle" (Harvey, 2003: 26). If we historicise the interaction between humans and water resources, in general terms, we can argue that hydraulic infrastructures are a familiar, reassuring, and reliable (the Dujiangyan irrigation scheme in China, for instance, has already served for twenty-two centuries) way to develop societies, and this is of course unsurprising given the inextricable link between water and the creation of ancient civilisations.[3] And indeed, who could possibly argue against the usefulness and necessity of having aqueducts, mills, and sewage disposal and irrigation systems? Furthermore, we can contend that the technology behind hydraulic infrastructures (with the exception of desalination and water purification techniques) did not lead to any major breakthrough since the industrial revolution in the nineteenth century.

If we take, for instance, the case of hydropower (see in this book chapters 7, 11, and 13), which towards the 1980s and 1990s had become internationally controversial, we can observe how the industry in the early 2000s has re-established its dominance as the main renewable energy source globally, particularly in emerging markets and less developed countries. As of 2015, 76% of all renewable electricity comes from hydropower plants, and the industry is booming (World Energy Council, 2015). Rather than being a result of technological advancement (most hydropower plants are still based on the Francis turbine, which was developed in 1849 by engineer James Francis; IHA, no date), this became possible through the incorporation of various sets of ideas (such as those related to sustainability, the water-energy nexus, scarcity, and integrated approaches to natural resources management) into the discursive frames of the hydropower sector, thus sidelining alternative readings of how to address particular societal and environmental challenges.

Linton (2010) eloquently argued in favour of seeing water as a process, something that is socio-materially produced and constantly renegotiated. This clearly resonates with Marx's central notion that "just as society produces man as man, so is society produced by him" (Marx, 1973: 37), but can also be connected to Gramsci's philosophy of praxis,[4] which the Sardinian defined as "absolute 'historicism', the absolute secularisation and earthliness of thought, an absolute humanism of history" (Gramsci, 1975: Q11, §27). This brings to surface the contradictions of contemporary societies, whereby men, who are seen as active processes, change themselves, other men, and the natural world, through their activities ("*per mezzo del lavoro e della tecnica*", Gramsci, 1975: Q10, §54). The intellectual life cannot be disjointed from the active life and men cannot be detached from nature (Gramsci, 1975: Q10, §37). And nevertheless, these activities (and consciousness) are the result of past processes, raising questions, therefore, on how we can actually achieve radical change if the past is reproduced in the present. The answer is, as Gramsci observed (1975: Q10, §54), that "the individual can associate himself with all the other individuals who want the same change, and if this change is rational, the individual can be multiplied for an impressive amount of times and can obtain a change which is by far more radical than what initially seemed possible". If we apply this perspective to the intricate relationship between humans and water resources, it seems indeed that the past is reverberated in the present. Water continues to be commodified and used for the benefit of the few and the reproduction of capital, and the creation of this collective consciousness that could lead to radical change in the way in which we share this natural resource is something that has yet to happen. This does not mean, however, that change is not happening or that contestations are not taking place, but this seems to be limited to isolated cases rather than leading to a paradigm shift. For instance, water privatisation (or dispossession) projects have been, at times, successfully challenged and reversed (Swyngedouw, 2005), with the most recent instance being the October 2017 decision of Indonesia's Supreme Court to restore public water services in Jakarta, since private companies "failed to protect" the citizens' right to water (Human Rights Watch, 2017). And yet, the global water privatisation agenda is not losing steam as the world's largest development institution, the World Bank, continues to push for privatisation as a key solution to the global water crisis (The World Bank, 2016).

The above also raises a series of questions stemming from the materiality of water (see, among others, Grundy-Warr et al., 2015; Steinberg and Peters, 2015; Lavau, 2013) and its political implications. H_2O as a chemical substance has indeed a wide range of effects on society and social relations. These can be very visible and relate, for instance, to its abundance or scarcity (floods or droughts), but can also be less noticeable, at least in the short term, and yet no less significant. This is the case of contaminated or polluted water, which is used as a drinking water source by at least 2 billion people, transmitting diseases such as cholera and typhoid (WHO, 2017), and whose effects were most strikingly visible in 2017 Yemen's cholera outbreak. As Bakker (2009: 517) argued, scholars (and geographers in particular) have to walk a thin line "between incorporating materiality . . . and avoiding the spectre of environmental determinism". This implies that the role of human agency should not be downplayed by deterministic considerations on the materiality of water, and yet, the role of the latter in shaping human societies should not be underestimated either. Besides its symbolic and cultural value, H_2O is essential to industry, urbanisation, and agriculture, and yet it is also subject to pollution and variability in its flows, and this of course provides a challenge (and many frustrations) to policymakers that desire to control it and to water governance experts more generally (Bakker, 2012). The more-than-human is also deeply enmeshed with water and with hydraulic infrastructures, as Mitchell (2002), for instance, tellingly illustrated in the case of the proliferation of the anopheles mosquito following the construction of the Aswan High Dam in Egypt. These dimensions of water are clearly linked with the nation-state and the processes through which it is formed and

10 *Filippo Menga and Erik Swyngedouw*

contested over time. Furthermore, and as the ongoing Flint water crisis in the United States clearly illustrates (Sadler and Highsmith, 2016), water can provide an explanatory tool for economic segregation and inequalities both in the Global North and in the Global South.

Water, technology, and the nation–state

The chapters in this volume engage with the above themes, shedding light on the often intangible and intricate web of relationships linking water, technology, and the nation–state, and attempting to make sense of pressing hydrosocial matters in the contemporary world. They do so through a range of both empirically grounded and theoretical critical work which covers four continents: North America, Europe, Africa, and Asia. All chapters are interconnected, even though they consider different conceptual and theoretical underpinnings, and some of the chapters speak to one another explicitly, both theoretically and, more evidently, geographically. Geography is indeed the arbitrary criteria that we have adopted to organise this volume, where the chapters are therefore loosely ordered from West to East.

In chapter 2, Joe Williams delineates the historical emergence of seawater desalination in the Colorado River, which is shared by the United States and Mexico. Williams applies an innovative theoretical perspective that brings together political ecology, assemblage theory, critical water geography, and transboundary water studies. This serves to illustrate the contradictions of capital. He argues that desalination represents a techno-political strategy and, ultimately, a technological fix to longstanding conflicts and tensions related to the governance of the Colorado River. The chapter, which sketches the historical development of the lower Colorado River Basin from the 1950s until the present time, provides an alternative reading of the challenges stemming from transboundary water governance, while also offering critical insight into the 'next big thing' when it comes to large-scale water technologies, namely desalination.

We then move to Europe with chapter 3, in which Santiago Gorostiza, Hug March, and David Saurí examine the ascendancy and dismissal of one the most ambitious and peculiar hydraulic infrastructures recently put forward in Europe (matched perhaps only by the proposed Strait of Messina Bridge in Italy): the Rhône Water Transfer Project between the Spanish region of Catalonia and the French region of Languedoc-Roussillon. We have indeed pointed out the conceptual fuzziness attached to the notion of nation-state, and this chapter fittingly positions this proposed canal within the broader context of surging Catalan secessionism as a challenge to Spanish political centralism. The authors illustrate how the Rhône Water Transfer Project – an infrastructure whose actual realisation has remained largely hypothetical – emerges as a discursive construction imbricated with pro-European ideas, but which incorporated over time emerging discourses such as the one on climate change.

This resonates, in part, with chapter 4, which brings us to Southern France. In this chapter, Jamie Linton and Etienne Delay take the case study of the Vinça Dam to illustrate how the centralised French state used a dam as a means to gain territorial presence in a region, the Eastern Pyrénées, which has been historically resistant to its control. This has been possible, the authors argue, through the shift from gravity to pressurised irrigation and the consequent renegotiation of the social relations between farmers and technocrats. Starting in the 1970s, the French state used the Vinça Dam to begin producing 'modern water' and offer hydrological certainty through technology, ultimately leading to the weakening of the local social structures and the commodification of water resources in the region. As the authors observe, this corroborates recent research (Zeitoun et al., 2016) questioning dominant approaches to water security understood in terms of certainty of water flows, pointing out that more reliable water supplies do not necessarily bring benefit to farmers.

In chapter 5, Emanuele Fantini, Tesfaye Muluneh, and Hermen Smit take the case study of the Beles Sugar Development Project in Ethiopia, a large-scale irrigation project funded by the state-owned Ethiopian Sugar Corporation (ESC), to discuss the interplay between state-building, water management, and large-scale land acquisitions. Ethiopia enacts its objective of being a developmental state through the practice of territorialisation, which is epitomised by the transformation and restructuring of a peripheral part of the country. Through a remarkable and varied amount of data sourced from fieldwork, the authors outline a wide and complex array of dynamics that are leading to the resettlement of peoples and the redistribution of resources and labour. The Beles Sugar Development Project also emphasises the contradictory process through which contemporary Ethiopia tries to mediate between being both a federal republic and an authoritarian state.

Chapter 6, by Ramy Hanna and Jeremy Allouche, focuses on one of Ethiopia's historical rivals for the use of the waters of the Nile River, Egypt. The authors build on previous work carried out by Allouche (2005) to introduce the concept of 'water nationalism', which serves to narrate the overlapping processes of state-building and nation-making in modern Egypt. Hanna and Allouche shed light on the strategies that the Egyptian ruling elites adopted to enrol water in the top-down enactment of their hydraulic mission and how this nurtured the dissemination of a particular idea of the nation. Yet, and in line with the argument put forward by Menga (2016, 2018), the domestic dimension is intimately interrelated with the international one, and this challenges the successful formation of the Egyptian entrepreneurial state.

In Chapter 7, Bengi Akbulut, Fikret Adaman, and Murat Arsel provide an insightful overview of the hydropower sector in Turkey, a country that in recent years has launched several large hydropower projects including the 'Southeastern Anatolia Project' (also known as the GAP). Yet, rather than focusing on large dams, as it is usually the case for literature on the country, the authors offer a novel reading by examining the social conflicts and the widespread opposition to smaller hydropower plants in North-eastern Anatolia during the last decade.

12 *Filippo Menga and Erik Swyngedouw*

Here, a Gramscian lens serves to position hydropower interventions within the broader setting of state–society relationships, and this illuminates the mechanisms through which the developmental state seeks consent from society.

Chapter 8 has Panayiota Pyla and Petros Phokaides exploring the recent history of water management in Cyprus, an island marked by droughts, severe water shortages, and conflicts over water allocation between different stakeholders. Taking as a case study the strategy adopted by the UN in the 1960s to improve water management in the Republic of Cyprus, the authors extricate the complex interrelation between water infrastructures and internal politics, placing this in the broader setting of the geopolitical tensions between Britain, Greece, and Turkey over the so-called 'Cyprus problem'. This serves to advance a critique of a series of initiatives that attempt to use water as a tool to solve the ongoing Cyprus dispute, overlooking, however, the historical complexities attached to water politics in the island.

In Chapter 9, Muna Dajani and Michael Mason focus on what is usually referred to as one the main water conflict hotspots globally, the occupied Golan Heights that Israel seized from Syria in 1967, thus asserting its monopoly of control over its water resources. The authors employ colonial theory and the concept of hydrosocial territories to provide an account of the ways in which the local Syrian population, the Jawlany, is contesting Israel's use of water technologies as a means to assert state territorialisation. The empirical evidence presented shows that the Jawlany are doing so through an ingenious counter-strategy of de-territorialisation in which water infrastructures, such as rainwater reservoirs and parallel pipelines, have been designed and used to bypass the discriminatory restrictions on the allocation of water for agricultural uses imposed by the Israeli settlers.

Chapter 10, authored by Andrea Zinzani, questions the effectiveness of development initiatives in the Talas River waterscape shared by the Central Asian countries of Kyrgyzstan and Kazakhstan. Zinzani advances the notion of the Conflicting Borderlands Hydrosocial Cycle to explore how the multiscalar complexities and the institutional restructuring that took place during the last decade have effectively hindered the work of the Chu-Talas Commission, an organisation that is generally considered as a success story for water cooperation in the region. In particular, the author illustrates how the institutional recentralisation and the shift in infrastructural property regimes in Kazakhstan played a crucial role in redefining power relations between Kyrgyzstan and Kazakhstan, which ultimately resulted in a benefit for the latter.

In Chapter 11 Austin Lord takes a close look at Nepal, an understudied South Asian country that in recent years has embarked upon an impressive number of hydropower projects. Lord adopts the notion of hydroscape to analyse the large investments in the hydropower sector that occurred after the devastating earthquakes that hit Nepal in April and May of 2015. His empirically rich study provides an account of Nepal's aspirations of becoming a hydropower nation, in what emerges as a complex and emotional rhetoric grounded on the narratives of energy security and energy sovereignty. This highly speculative development plan, however, clashes with Nepal's intense seismic activity. The author concludes on a gloomy note, reminding us that while humans might forget about nature, nature, in turn, might sooner or later crush these short-sighted hopes of national resurgence through hydropower.

In chapter 12, David J.H. Blake contributes to the literature on hydrocracies and hydraulic societies through an examination of ambitious irrigation plans with a six-decade history and recently reinvigorated by the incumbent military regime for Northeast Thailand. Blake advances the concept of *irrigationalism* to underline the ideology formed by the inextricable connection between irrigation developmentalism and top-down attempts of human domination over nature and society. Such efforts, the author argues, overlap with an elite-driven project aimed at propagating an idea of 'Thainess' embedded in romantic reconstructions of an idealised past, whereby irrigated agriculture becomes a state-sanctioned activity necessary to preserve Thai culture and promote notions of nationalism. This process, however, happens to the detriment of more viable alternatives for Thailand's development path.

In the final chapter, Covell F. Meyskens provides a historical account of the genesis of one of the largest and most discussed dams built in recent times, the Three Gorges Dam in China. As Meyskens explains, Western imperialist pressures in the mid-nineteenth century triggered a new understanding of technology that has been appropriated by subsequent Chinese leaders in the early twentieth century. This, in turn, led to different technological styles and approaches to management, whereby Sun Yat-sen's technocratic impetus is replaced by Mao's discursive emphasis on mass mobilisation and national voluntarism. Yet, Meyskens argues, such a faith on popular mobilisation came at the expenses of technical expertise, with serious consequences for the first attempts to build this dam.

The above discussion, together with the insights provided by the chapters forming this book, sheds light on the interconnections and mutual ramifications between water, technology, and the nation-state, as they emerged over time and across scale and place. In light of the resurgence of the hydropower sector, the persistence of inequalities and the increasingly challenging and upsetting state of global water governance, this volume provides background and evidence aimed at addressing some pressing questions. What clearly emerges, we argue, is that this book should not be read as a book about water, or at least, not only about water. Rather, we contend that water provides an excellent lens through which some of the contradictory and often unequal dynamics that shape social interactions can be interpreted and explained. Water helps us understand how particular spatial scales can be produced, but also how they can be contested, and this can be transferred and applied to other settings where different forces and interests are at work. The chapters are indeed linked by a common thread, and this was a deliberate choice: all of the case studies examined present, in different forms, choreographies of oppression and domination, where the interests of water users are being obscured by broader power dynamics that are not necessarily related to water. Yet, we show that water also generates strategies of

14   *Filippo Menga and Erik Swyngedouw*

resistance and contestation, and these, even though they are often scattered, have the potential to be channelled into the creation of a collective consciousness that could lead to the radical change needed to share our resources more equally.

Notes

1 See Antonsich (2009) for a comprehensive analysis of the issues arising from the increasing convergence between nation and state, a phenomenon that he calls 'the crisis of the hyphen'.
2 As a result of the growing opposition to large dams, in 1997 the World Bank launched the work of the WCD, a body tasked to review the development effectiveness of large dams, along with their social, economic, and environmental impact. This seeming new era for the hydropower sector was also marked by the establishment of the International Hydropower Association (IHA), an international organisation created under the auspices of UNESCO in 1995. In 2011, the IHA published the Hydropower Sustainability Assessment Protocol (accessible at this link: www.hydrosustainability.org/IHAHydro4Life/media/PDFs/Protocol/ hydropower-sustainability-assessment-protocol_web.pdf), a document containing an elaborate complex scorecard to rate the sustainability of dam projects.
3 For a detailed historical account of the link between water and civilisation refer to Yevjevich (1992).
4 For an excellent theoretical discussion of Gramsci's philosophy of praxis, see Loftus (2013).

References

Agamben, G. (2005). *State of Exception.* Chicago: The University of Chicago Press.
Agnew, J. (1994). The territorial trap: The geographical assumptions of international relations theory. *Review of International Political Economy*, 1(1), pp. 53–80.
Agnew, J. (2010). Still trapped in territory? *Geopolitics*, 15(4), pp. 779–784.
Ahlers, R., Cleaver, F., Rusca, M. and Schwartz, K. (2014). Informal space in the urban waterscape: Disaggregation and co-production of water services. *Water Alternatives*, 7(1), pp. 1–14.
Akhter, M. (2015). The hydropolitical cold war: The Indus waters treaty and state formation in Pakistan. *Political Geography*, 46, pp. 65–75.
Allouche, J. (2005). *Water Nationalism: An Explanation of the Past and Present Conflicts in Central Asia, the Middle East and the Indian Subcontinent?* (Doctoral dissertation, Institut universitaire de hautes études internationales).
Anderson, B. (2006). *Imagined Communities.* London and New York: Verso.
Antonsich, M. (2009). On territory, the nation-state and the crisis of the hyphen. *Progress in Human Geography*, 33(6), pp. 789–806.
Bakker, K. (2009). Water. In Castree, N., Demeritt, D., Liverman, D. and Rhoads, B. (Eds.), *A Companion to Environmental Geography*, pp. 515–532. Chicester: Wiley-Blackwell.
Bakker, K. (2012). Water: Political, biopolitical, material. *Social Studies of Science*, 42(4), pp. 616–623.
Blaikie, P. and Brookfield, H. (Eds.). (1987). *Land Degradation and Society.* Abingdon: Routledge.
Boelens, R., Hoogesteger, J., Swyngedouw, E., Vos, J. and Wester, P. (2016). Hydrosocial territories: A political ecology perspective. *Water International*, 41(1), pp. 1–14.
Brown, J.C. and Purcell, M. (2005). There's nothing inherent about scale: Political ecology, the local trap, and the politics of development in the Brazilian Amazon. *Geoforum*, 36(5), pp. 607–624.

Budds, J. and Hinojosa, L. (2012). Restructuring and rescaling water governance in mining contexts: The co-production of waterscapes in Peru. *Water Alternatives*, 5(1), pp. 119–137.

Castree, N. (2008). Neoliberalising nature: The logics of deregulation and reregulation. *Environment and Planning A*, 40(1), pp. 131–152.

Castree, N. (2014). *Making Sense of Nature*. Abingdon: Routledge.

Chesterman, S. (2004). *You, the People: The United Nations, Transitional Administration, and State-building*. Oxford: Oxford University Press.

Cox, K.R. (1998). Spaces of dependence, spaces of engagement and the politics of scale, or: Looking for local politics. *Political Geography*, 17(1), pp. 1–23.

Delaney, D. and Leitner, H. (1997). The political construction of scale. *Political Geography*, 16(2), pp. 93–97.

Dittmer, J. (2014). Geopolitical assemblages and complexity. *Progress in Human Geography*, 38(3), pp. 385–401.

Flint, C. and Taylor, P.J. (2007). *Political Geography: World-economy, Nation-state, and Locality*. New York: Pearson Education.

Furlong, K. (2006). Hidden theories, troubled waters: International relations, the 'territorial trap', and the Southern African Development Community's transboundary waters. *Political Geography*, 25(4), pp. 438–458.

Gandy, M. (2003). *Concrete and Clay: Reworking Nature in New York City*. Cambridge: MIT Press.

Giddens, A. (1981). *Contemporary Critique of Historical Materialism, Volume 1: Power, Property and the State*. Cambridge: Polity Press.

Giddens, A. (1994). *Beyond Left and Right*. Cambridge: Polity Press.

Gramsci, A. (1975). *Quaderni del carcere*. Torino: Einaudi.

Grundy-Warr, C., Sithirith, M. and Li, Y.M. (2015). Volumes, fluidity and flows: Rethinking the 'nature' of political geography. *Political Geography*, 45, pp. 93–95.

Gupta, J. and Pahl-Wostl, C. (2013). Global water governance in the context of global and multilevel governance: Its need, form, and challenges. *Ecology and Society*, 18(4), p. 53.

Habermas, J. (1970). *Toward a Rational Society*. Boston: Beacon Press.

Harris, L.M. (2005). Navigating uncertain waters. In Flint, C. (Ed.), *The Geography of War and Peace: From Death Camps to Diplomats*, pp. 259–279. New York: Oxford University Press.

Harris, L.M. and Alatout, S. (2010). Negotiating hydro-scales, forging states: Comparison of the upper Tigris/Euphrates and Jordan River basins. *Political Geography*, 29(3), pp. 148–156.

Harvey, D. (2003). *The New Imperialism*. Oxford: Oxford University Press.

Herod, A. and Wright, M.W. (Eds.). (2008). *Geographies of Power: Placing Scale*. Oxford: John Wiley & Sons.

Heynen, N.C., Kaika, M. and Swyngedouw, E. (Eds.). (2006). *In the Nature of Cities: Urban Political Ecology and the Politics of Urban Metabolism*. Abingdon: Routledge.

Horkheimer, M. (1974). *Eclipse of Reason* (Vol. 1). New York: Bloomsbury Publishing.

Human Rights Watch. (2017). *Indonesia's Supreme Court Upholds Water Rights*. Available from: www.hrw.org/news/2017/10/12/indonesias-supreme-court-upholds-water-rights [Accessed 3 May 2017].

IHA (International Hydropower Association). (no date). *A Brief History of Hydropower*. Available from: www.hydropower.org/a-brief-history-of-hydropower [Accessed 6 May 2017].

Innes, J.E., Connick, S. and Booher, D. (2007). Informality as a planning strategy: Collaborative water management in the CALFED Bay-Delta Program. *Journal of the American Planning Association*, 73(2), pp. 195–210.

IPCC (Intergovernmental Panel on Climate Change). (2012). *Managing the Risks of Extreme Events and Disasters to Advance Climate Change Adaptation*. New York: Cambridge University Press.

Kaika, M. (2005). *City of Flows: Modernity, Nature, and the City.* Abingdon: Routledge.

Kuus, M. and Agnew, J. (2008). Theorizing the state geographically: Sovereignty, subjectivity, territoriality. In Cox, K., Robinson, J. and Low, M. (Eds.), *The Handbook of Political Geography*, pp. 117–132. London: Sage.

Lavau, S. (2013). Going with the flow: Sustainable water management as ontological cleaving. *Environment and Planning D: Society and Space*, 31, pp. 416–433.

Linton, J. (2010). *What Is Water? : The History of a Modern Abstraction.* Toronto: UBC Press.

Linton, J. and Delay, E. (2018). Death by certainty: The Vinça dam, the French state, and the changing social relations of irrigation the Têt basin of the Eastern French Pyrénées. In Menga, F. and Swyngedouw, E. (Eds.), *Water, Techonology and the Nation-State* (in press). Abingdon: Routledge Earthscan.

Loftus, A. (2009). Rethinking political ecologies of water. *Third World Quarterly*, 30(5), pp. 953–968.

Loftus, A. (2013). Gramsci, nature, and the philosophy of praxis. In Loftus, A., Michael, E., Gillian, H., Stefan, K. and Alex, L. (Eds.), *Gramsci: Space, Nature, Politics*, pp. 178–196. Oxford: John Wiley & Sons.

Marx, K. (1973). *The Economic and Philosophic Manuscripts of 1844.* London: Lawrence and Wishart.

Mehta, L., Veldwisch, G. J. and Franco, J. (2012). Introduction to the special issue: Water grabbing? Focus on the (re) appropriation of finite water resources. *Water Alternatives*, 5(2), pp. 193–207.

Menga, F. (2015). Building a nation through a dam: The case of Rogun in Tajikistan. *Nationalities Papers*, 43(3), pp. 479–494.

Menga, F. (2016). Domestic and international dimensions of trans-boundary water politics. *Water Alternatives*, 9(3), pp. 704–723.

Menga, F. (2017). Hydropolis: Reinterpreting the polis in water politics. *Political Geography*, 60, pp. 100–109.

Menga, F. (2018). *Power and Water in Central Asia.* Abingdon: Routledge.

Mitchell, T. (2002). *Rule of Experts: Egypt, Techno-politics, Modernity.* Berkeley: University of California Press.

Mollinga, P.P. (2008). Water, politics and development: Framing a political sociology of water resources management. *Water Alternatives*, 1(1), pp. 7–23.

Mouffe, C. (2005). *On the Political.* London: Routledge.

Neumann, R.P. (2009). Political ecology: Theorizing scale. *Progress in Human Geography*, 33(3), pp. 398–406.

Newman, D. and Paasi, A. (1998). Fences and neighbours in the postmodern world: Boundary narratives in political geography. *Progress in Human Geography*, 22(2), pp. 186–207.

Norman, E.S. (2014). *Governing Transboundary Waters: Canada, the United States, and Indigenous Communities.* Abingdon: Routledge.

Norman, E.S. and Bakker, K. (2009). Transgressing scales: Water governance across the Canada – US borderland. *Annals of the Association of American Geographers*, 99(1), pp. 99–117.

Norman, E.S., Bakker, K. and Cook, C. (2012). Introduction to the themed section: Water governance and the politics of scale. *Water Alternatives*, 5(1), pp. 52–61.

Norman, E.S., Cook, C. and Cohen, A. (Eds.). (2015). *Negotiating Water Governance: Why the Politics of Scale Matter.* Farnham: Ashgate.

Obertreis, J., Moss, T., Mollinga, P. and Bichsel, C. (2016). Water, infrastructure and political rule: Introduction to the special issue. *Water Alternatives*, 9(2), pp. 168–181.

OECD (Organisation for Economic Co-operation and Development). (no date). *Water Withdrawals.* Available from: https://data.oecd.org/water/water-withdrawals.htm [Accessed 6 September 2017].

Oxfam. (2017). *An Economy for the 99%.* Available from: www.oxfam.org/en/research/economy-99 [Accessed 11 November 2017].

Painter, J. (2006). Prosaic geographies of stateness. *Political Geography*, 25(7), pp. 752–774.

Perreault, T. (2014). What kind of governance for what kind of equity? Towards a theorization of justice in water governance. *Water International*, 39(2), pp. 233–245.

Rogers, S., Barnett, J., Webber, M., Finlayson, B. and Wang, M. (2016). Governmentality and the conduct of water: China's South – North Water Transfer Project. *Transactions of the Institute of British Geographers*, 41(4), pp. 429–441.

Roy, A. (2009). Why India cannot plan its cities: Informality, insurgence and the idiom of urbanization. *Planning Theory*, 8(1), pp. 76–87.

Sadler, R.C. and Highsmith, A.R. (2016). Rethinking Tiebout: The contribution of political fragmentation and racial/economic segregation to the flint water crisis. *Environmental Justice*, 9(5), pp. 143–151.

Sharp, L. (2017). *Reconnecting People and Water: Public Engagement and Sustainable Urban Water Management.* Abingdon: Routledge.

Smith, N. (2006). *There's No Such Thing as a Natural Disaster. Understanding Katrina: Perspectives From the Social Sciences.* Available from: http://understandingkatrina.ssrc.org/Smith [Accessed 17 October 2017].

Sneddon, C. and Fox, C. (2006). Rethinking transboundary waters: A critical hydropolitics of the Mekong basin. *Political Geography*, 25(2), pp. 181–202.

Sneddon, C. and Fox, C. (2015). A Genealogy of the Basin: Scalar Politics and Identity in the Mekong River Basin. In Norman, E.S., Cook, C. and Cohen, A. (Eds.), *Negotiating Water Governance: Why the Politics of Scale Matter*, pp. 41–57. Farnham: Ashgate.

Steinberg, P. and Peters, K. (2015). Wet ontologies, fluid spaces: Giving depth to volume through oceanic thinking. *Environment and Planning D: Society and Space*, 33, pp. 247–264.

Swyngedouw, E. (1997). Power, nature, and the city: The conquest of water and the political ecology of urbanization in Guayaquil, Ecuador: 1880–1990. *Environment and Planning A*, 29(2), pp. 311–332.

Swyngedouw, E. (2005). Dispossessing H2O: The contested terrain of water privatization. *Capitalism Nature Socialism*, 16(1), pp. 81–98.

Swyngedouw, E. (2006). *Power, Water and Money: Exploring the Nexus.* United Nations Human Development Report. Occasional Paper 2006/14.

Swyngedouw, E. (2010a). Impossible sustainability and the post-political condition. In *Making Strategies in Spatial Planning*, pp. 185–205, Dordrecht: Springer.

Swyngedouw, E. (2010b). *Place, Nature and the Question of Scale: Interrogating the Production of Nature.* Berlin, Germany: Brandenburgische Akademie der Wissenschaften.

Swyngedouw, E. (2015). *Liquid Power: Contested Hydro-modernities in Twentieth-century Spain.* Cambridge: MIT Press.

Swyngedouw, E. and Heynen, N.C. (2003). Urban political ecology, justice and the politics of scale. *Antipode*, 35(5), pp. 898–918.

WCD (World Commission on Dams). (2000). *Dams and Development: A New Framework for Decision-making: The Report of the World Commission on Dams.* Abingdon: Earthscan.

Weber, M. (1978). *Economy and Society: An Outline of Interpretive Sociology* (Vol. 1). Berkeley: University of California Press.

White, R. (1995). *The Organic Machine: The Remaking of the Columbia River.* New York: Hill and Wang.

WHO (World Health Organization). (2015). *Progress on Sanitation and Drinking Water – 2015 Update and MDG Assessment.* New York: UNICEF and World Health Organization.

WHO (World Health Organization). (2017). *Drinking Water.* Available from: www.who.int/mediacentre/factsheets/fs391/en/ [Accessed 16 September 2017].

Williams, J., Bouzarovski, S. and Swyngedouw, E. (2014). *Politicising the Nexus: Nexus Technologies, Urban Circulation and the Coproduction of Water – Energy.* Nexus Network Think Piece Series, Paper, 1.

Wilson, J. and Swyngedouw, E. (2014). *The Post-political and Its Discontents: Spaces of Depoliticization, Spectres of Radical Politics.* Edinburgh: Edinburgh University Press.

Wittfogel, K.A. (1957). *Oriental Despotism: A Comparative Study of Total Power.* Binghamton: Yale University Press.

The World Bank. (2016). *FAQ – World Bank Group Support for Water and Sanitation Solutions.* Available from: www.worldbank.org/en/topic/water/brief/working-with-public-private-sectors-to-increase-water-sanitation-access [Accessed 17 September 2017].

World Energy Council. (2015). *World Energy Resources Charting the Upsurge in Hydropower Development 2015.* London: World Energy Council.

Worster, D. (1984). The Hoover Dam: A study in domination. *The Social and Environmental Effects of Large Dams*, 2, pp. 17–24.

WWAP (United Nations World Water Assessment Programme). (2016). *The United Nations World Water Development Report 2016.* Water and Jobs. Paris: UNESCO.

WWAP (United Nations World Water Assessment Programme). (2017). *The United Nations World Water Development Report 2017. Wastewater: The Untapped Resource.* Paris: UNESCO.

Yevjevich, V. (1992). Water and civilization. *Water International*, 17(4), pp. 163–171.

Zeitoun, M., Lankford, B., Kreuger, T., Forsyth, T., Carter, R., Hoekstra, A.Y., Taylor, R., Varis, O., Cleaver, F., Boelens, R., Swatuk, L., Tickner, D., Scott, C.A., Mirumachi, N. and Matthews, N. (2016). Reductionist and integrative approaches to complex water security challenges. *Global Environmental Change*, 39, pp. 143–154.

Zinzani, A. and Menga, F. (2017). The Circle of Hydro-Hegemony between riparian states, development policies and borderlands: Evidence from the Talas waterscape (Kyrgyzstan-Kazakhstan). *Geoforum*, 85, pp. 112–121.

Žižek, S. (2000). *The Ticklish Subject: The Absent Centre of Political Ontology.* New York: Verso.

2 The ocean bountiful?

De-salination, de-politicisation,
and binational water governance
on the Colorado River

Joe Williams

Introduction

In May 2010 an extraordinary document was published. Although technical in nature and understated in terms of its political implications, this document in many ways represented the culmination of nearly a century of disputes on the Colorado River. Eighty-eight years after the signing of the Colorado River Compact, under which the entire annual flow of the river was (over)allocated, and 66 years after the United States and Mexico reached an unstable compromise over their respective entitlements to water, a number of the basin's largest water users were looking for new ways to fix intensifying disputes and shortages. Their conclusion: manufacture water to add to the Colorado River Basin. Four water agencies in the United States and four from Mexico collaborated in a feasibility study proposing the construction of a large seawater desalination plant located about 30 km south of the border at Rosarito Beach (SDCWA, 2010). This international infrastructure project would be financed by agencies on the Colorado basin on both sides of the border. Desalted ocean water would then be transferred to the US either by pipeline (as 'wet water') or through the transfer of entitlements to river water (or 'dry water'). This way, water users as far inland as Las Vegas could finance coastal desalination in return for more secure access to Colorado water (Shrestha et al., 2011). In other words, by effectively increasing overall supply in the basin, 'binational' desalination was proposed as a technical fix for the intense political disputes that beleaguer water governance between the states on the Colorado River.

This chapter is about the historical emergence of seawater desalination as a techno-political strategy in the governance of international and multi-state waters in southwest United States and northwest Mexico. Transboundary and multi-state water governance is, of course, a topic of growing pertinence in policy, industry and business, and academia (e.g. Pahl-Wostl, 2015; Rieu-Clarke et al., 2015). This chapter traces the technological, discursive and political formations that have coalesced around desalination as a water governance 'solution' for the Colorado River states. From the utopian visions of the 1960s and 1970s that saw the combination of desalination and nuclear energy technologies as a path towards a resource-abundant future; to the use of purification technologies

20 *Joe Williams*

to fix specific governance problems in the 1980s and 1990s; and the more recent addition of seawater as a water supply diversification and decentralisation strategy, desalination has intersected water governance and international politics in extraordinary (and largely unexplored) ways for the last half-century. Although the political ecology and political economy of desalination has recently become the focus of critical scholarship concerning national or regional hydro-development (Loftus and March, 2016; March et al., 2014; McEvoy, 2014; Swyngedouw, 2013; Swyngedouw and Williams, 2016; Feitelson and Rosenthal, 2012), with some notable exceptions (Wilder et al., 2016; Aviram et al., 2014), there has been very little consideration of desalination as an important political technology of transboundary water governance.

The chapter argues that desalination, first and foremost, represents a technological fix for the dialectically intertwined challenges of politically contested terrestrial water governance and economic development facing the Colorado River states. The use of this language invokes two related notions of the technical fix – first, as a spatial fix for internal contradictions of capital. Drawing on the work of David Harvey (1996, 2006, 2014), Neil Smith (1984), and more recently Micheal Ekers and Scott Prudham (2015), and Jason Moore (2015), I argue that the primary function of desalination is to secure the socio-ecological conditions (i.e. the security and resilience of water supply) necessary for the continued expansion of capital accumulation and the economic development of the lower Colorado River Basin states. Second, drawing on the growing body of literature on the ecological conditions of post-politics (Kenis and Lievens, 2014; Wilson and Swyngedouw, 2014; Swyngedouw, 2011), the chapter argues that desalination is emerging as a political fix for contested relations of water governance. Mobilising Murray Li's (2011) concept of the process of 'rendering society technical,' I argue that the development of seawater desalination for the Colorado River states represents an attempt to secure reliable water supply and fix the political problems of allocation without addressing the underlying causes of those problems.

The chapter proceeds in four phases. It begins with a brief overview of desalination as a technology of water governance. The chapter then sketches the historical development of the lower Colorado River Basin and the origins of contestations between its constituent states, before considering the emergence of desalination, firstly as an elusive panacea, and more recently as a significant technology in the scalar restructuring of water governance on the Colorado.

Water, desalination and the nation state

Although the drivers of desalination are always highly contextual – the governance decisions that led to desalination developments in Rosarito are, for example, very different to those in Riyadh – proponents laud the desalination 'solution' for producing a rainfall and climate-independent source of water to address the combined challenges of increasing demand and reducing traditional supply. "Seawater desalination," in a word, "offers a seemingly unlimited, steady supply of high-quality water, without impairing natural freshwater ecosystems" (Elimelech and Phillip, 2011: 713). There are two broad methods of removing dissolved impurities from saltwater: thermal distillation and membrane. Thermal distillation, which, put simply, involves the separation of salt from water through the creation of water vapour, can be achieved through a number of processes (Khawaji et al., 2008). Until the 1990s, a process called multi-stage flash distillation was the industry standard technology for municipal-scale desalination plants, and remains the most widely used technology in the Middle East (Al-Kharaghouli and Kazmerski, 2013).

Membrane desalination processes, by contrast, separate non-saline water from a saline brine reject with a physical barrier. Techniques include electrodialysis, membrane distillation, forward osmosis and reverse osmosis. Reverse osmosis (RO), where saline water is forced at high pressure through membranes that trap dissolved salt and allow pure water to pass through, is now the most dominant desalting technique globally (Fritzmann et al., 2007). Improvements in membrane technology and the introduction of energy recovery devices have reduced the energy consumption of seawater conversion from 20 kWh/m³ in the 1970s to 2 kWh/m³ today (Peñate and García-Rodríguez, 2012). This has, in part, facilitated a *global* profusion of large-scale seawater desalination plants in cities as diverse as Singapore, San Diego, London, Tel Aviv, Melbourne and Alicante – a phenomenon that has accelerated only over the last 10–20 years (Feitelson and Jones, 2014).

Water desalination has, in just a few short decades, undergone an extraordinary transformation from fringe water source utilised only under the most extreme circumstances or for specific manufacturing functions, to a global industry, increasingly the focus of techno-managerial solutions to urban water stress. Although long overlooked in geography and the social sciences, the desalting phenomenon has lately attracted more sustained critical attention (March, 2015).

The contested Colorado

The Colorado River, sometimes referred to in the West as the 'river of law,' is one of the most contested, legislated and litigated river basins in the world (Reisner, 1986). From its source in the Rocky Mountains, the 'American Nile' flows through seven of the United States (Colorado, New Mexico, Utah and Wyoming in the Upper Basin and Arizona, California and Nevada in the Lower Basin) and the two Mexican states of Baja California and Sonora. The Colorado is not so much noted for its size than for the seasonality of its flow. Prone to seasonal flooding, before its comprehensive impoundment by the Bureau of Reclamation, streamflow for 10 months of the year would be only around 10% of peak flow during the spring. By the end of the 1920s, following the signing of the Colorado River Compact in 1922 and the Boulder Canyon Act of 1928, the entire annual streamflow of the river had been fully allocated between the Upper and Lower Basin states in the US. The history of the river since has been one of fierce disagreement and protracted legal battles over the fair apportionment of its waters between the two nations that make up its basin and the various states and stakeholders therein.

The disagreements between the United States and Mexico over allocations on the Colorado go back to the late 1920s. The United States had, by that time, fully allocated the river's annual output, and had begun work on the Boulder Dam (which would later become the Hoover Dam) and other works towards the comprehensive development of the lower basin. To this point, having been witnesses to this water-grabbing fever and eager to secure resources for economic development in Baja California and Sonora, in 1929 Mexican authorities argued that the farmable land area in the basin area amounted to 1.4 million acres. Mexico therefore claimed 5.5 million mega litres a year for irrigation, primarily in the Mexicali Valley, that should be guaranteed by the United States (Six States Committee, 1944). The US government, in turn, insisted that Mexico was not entitled to any more water than the average base-flow in dry months. The US offered to guarantee delivery of only 920,000 mega litres, the amount that Mexico had claimed for irrigation in 1928, before the signing of the Boulder Canyon Act and the full development of the lower basin by the Bureau of Reclamation.

Mexico had been unable to develop its own large projects on the Colorado for two main reasons. Firstly, given that only 3% of the basin is located in Mexico, by the time the river reaches the international border it is already flowing in a flat estuary. This means that geological and hydrological conditions in the entire Mexican portion are unsuitable for large dams, diversions and reservoirs. Any significant Mexican-led infrastructural projects would therefore have to be located within the United States. Secondly, Mexico is bound by the 1848 Treaty of Guadalupe-Hidalgo and the Gadsden Treaty of 1853 to ensure that the river remains navigable for US ships and trading vessels. Notwithstanding that navigation of the Lower Basin has been virtually impossible since the completion of works by the Bureau of Reclamation, given that now only rarely does any water from the Colorado actually reach the Gulf of California, these historic treaties clearly preclude any damming or significant diversions of the river by Mexico. For these reasons, Nevada Senator Key Pittman argued at a debate on the Boulder Canyon Project Act in 1928 that "the only water that Mexico could claim would be water that she has appropriated from the natural flow of the stream," and was therefore entitled to "none of the benefits of the water increased by our impounding works" (cited by McCarran, 1945: 50). By contrast, the Mexican government argued at the time, firstly, that the extensive development of the river by the US violated these same treaty agreements, and secondly, that any diversions of the Colorado by the US would reduce the total annual flow to Mexico, of which its citizens and farmers had a right to use fair proportion. Mexico was, therefore, entitled to a share of the annual yield from the Bureau's projects.

The concern around these claims and counter-claims, and their potential implications for economic development in the desert west on both sides of the border, remained an issue of severe political disagreement between the two nations. The California governor at the time, Earl Warren, even went so far as to argue that;

> Every acre in Mexico which is irrigated by Colorado River water, necessitates that a corresponding acre in this country be doomed forever to the sterility of desert.
>
> (Warren, 1945: 5)

Following Mexico's claim in 1929 to 5.5 million mega litres of Colorado water a year, the disputes continued unresolved for more than a decade. In this time Mexican off-takers expanded their combined annual use to 2.2 million mega litres. A resolution was in the development interests of both nations. From the Mexican perspective, a legally allocated apportionment of water would provide security for agricultural development and urban growth. At the same time, it was in the interests of off-takers in the United States – between whom the annual yield of the river was fully allocated but their capacity to withdraw was not yet fully developed – to, in a sense, cut their losses by limiting the withdrawals in Mexico that had grown rapidly in the absence of a treaty (CRWU Committee, 1945).

Finally, in 1944 the two countries reached a compromise and in 1945 ratified an agreement on the "Utilization of waters of Colorado and Tijuana Rivers and of the Rio Grande." Under this treaty the United States committed to delivering to Mexico 1.8 million mega litres a year of Colorado water suitable for irrigation. This compromise, however, was not well received by all water users. In California, the state which stood to lose the most from any allocation of water to Mexico, it was received with particular hostility. Representatives of off-takers argued that "any guarantee of water to Mexico must invade the commitments made by the United States to its own projects (Colorado River Board of California, 1944: 3). Although this agreement stands today, the United States has since consistently been in breach of its delivery commitment. The implications of this are discussed more in the next section.

This brief sketch of the historic disagreements over entitlements to Colorado water is given for two reasons. Firstly, illustrate that the governance of the Colorado River since the 1920s has been characterised by disputes between the basin's constituent states and right-holders. Secondly, that the infrastructural projects of the 1930s–1970s and associated governance structures entrenched a model of economic development based on large agribusiness and a highly inert and hierarchical system of water rights.

Desalination, the elusive catholicon

Since the mid-twentieth century the promise of saline water conversion, or desalination, has emerged in various guises as a potential panacea for the contested politics of the Colorado River. In the United States interest in desalination really began to pick up in the 1950s and 1960s, under direction from a

24   *Joe Williams*

well-funded federal programme. This began in 1952 when Congress authorised funding through the Department of the Interior, under the Water and Power Development scheme. Research was coordinated by the newly established Office of Saline Water, which operated between 1955 and 1974. The era of state-funded research and development (R&D) reached its height under the Kennedy Administration, when desalination was a high-priority issue for the government. When the country's first ocean desalting test plant – a Multi-Effect Vertical Distillation facility in Freeport, Texas, with a 1 million gallon per day capacity – opened in 1961, President Kennedy said:

> I can think of no cause and no work which is more important, not only to the people of this country, but to people all around the globe . . . I am hopeful that the United States will continue to exert great leadership in this field, and I want to assure the people of the world that we will make all the information that we have available to all people. We want to join with them, with the scientists and engineers of other countries in their efforts to achieve one of the great scientific breakthroughs of history.
>
> (Kennedy, 1961)

The desalination programme during this time was international in scope and ambition. The US government even signed an agreement with the USSR for the exchange of scientific information relating to saline water conversion. R&D funding was primarily contracted out to private research companies and engineering consultants. General Atomics, a subsidiary of General Dynamics, for instance, emerged as a major player. Its desalination group, Reverse Osmosis General Atomics (or ROGA), was very successful in attracting government funds and pioneered the reverse osmosis (RO) method of desalination, which now dominates the industry. Although desalination never became the panacea that Kennedy had envisioned, the programme was in many respects highly successful. It really kick-started the desalination industry and facilitated the development of novel technologies, like RO, that many in the industry believe would not have been developed otherwise.

During the 1960s desalination was the focus of interest at virtually every level of water management in the southwest United States from the federal level to the local water authority, and was seen by the state as a panacea for insoluble disagreements over access to the fully allocated riparian waters of the American west. For instance, the development of coastal desalting capacity was an important – although never realised – component of the federal government's efforts to resolve a series of protracted disputes between California and Arizona over allocations of Colorado water. The roots of the conflict went back to Arizona's refusal to ratify the 1922 Colorado River Compact because of the so-called 'tributary issue' over whether withdrawals from tributaries of the Colorado should be included in a state's overall allocation. After more than a decade of legal battles, a landmark ruling at the Supreme Court in 1963 effectively increased Arizona's allocation and therefore reduced availability for California.

A major planning document from the Department of the Interior (1963) – known as the Udall Plan for the Pacific Southwest – proposed the construction of a large desalter as a way of compensating Southern California for some of the water rights lost to Arizona. Similarly, throughout the 1960s and 1970s the Metropolitan Water District (MWD) of Southern California, the largest urban water wholesaler in the United States, also pursued collaborative desalination R&D programmes with its member agencies.

At the time, most of the proposed projects for large-scale desalination utilised the thermal distillation method – membrane technologies being still in the very early stages of development. Most were based around a model of co-production which is still utilised on the Arabian Peninsula, whereby desalination facilities are twinned with thermoelectric power plants to take advantage of waste heat (Kamal, 2005). During this time various proposals circulated for plants with capacities up to 570 mega litres a day – three times larger than the Carlsbad desalination plant, which began operating in 2015 and is currently the largest desalter in the Western hemisphere. In the United States in the 1970s, much as it was in Spain at the same time, the desalination panacea was premised on, and intimately linked with, the assumption of long-term cheap nuclear energy. This was, after all, the height of the atomic era, and it was generally agreed that large-scale desalination could take advantage of both low-cost energy and efficiencies through co-production with thermoelectric nuclear plants. Nuclear technology would, it was thought, solve the problems of energy and water supply simultaneously.

The huge efficiency gains borne out of the intensive R&D programmes during this time, combined with falling energy prices and the promise of unlimited cheap nuclear energy, led to optimistic (and entirely unrealistic) forecasts about the future cost of desalted water. Indeed, one respected Berkeley professor confidently calculated that because the unit cost had fallen so rapidly in the decade following the commencement of federal funding, by 1990 California would be able to produce desalted seawater at less than $30 a mega litre (Seckler, 1965). Even taking inflation into account, he could hardly have been more wrong: water is now sold from the recently completed Carlsbad desalting plant, which is touted as the most efficient of its kind in the world, at $1,900 a mega litre. Although several of the large facilities proposed in the 1960s and 1970s got past the feasibility and design stage (Holtom and Galstaun, 1965), as the nuclear sector went into decline and energy prices rose rather than fell, the desalination industry underwent corresponding changes, and the planned developments – including the large project proposed in the Udall Plan – were abandoned.

Unrealistic technological optimism aside, the plans for saltwater desalination during this period in many respects prefigured those of today. For instance, the Udall Plan and the MWD collaborative programmes clearly position desalination as a viable technological solution to increase the overall allocation in the Colorado River's lower basin. The contemporary binational desalting plans for Rosarito Beach effectively reproduce this logic, albeit through different technological, political and economic configurations. Even as early as the 1960s, then, ocean desalination was being proposed as a political 'solution' that addressed water supply issues without addressing those of a dysfunctional water rights system and a metabolic logic of capitalism based on agrarian accumulation in the context of an increasingly urbanising coastal economy. Rather than forming the centre of a new water paradigm, however, as was envisioned by some in the optimism of the 1960s, the few examples of successfully developed desalination plants in Southern California during this time were built as technological solutions to resolve specific inter-state political conflicts. Two examples stand out as particularly illuminating: the plant at Point Loma and the Yuma desalter.

In 1962 a company called Burns and Roe, funded by federal research grants, began operating a small seawater conversion plant at Point Loma, San Diego. This was a test facility, the second of five commissioned by the US government, designed to trial different techniques and produce the necessary data for the development of commercial facilities (Foster and Herlihy, 1965). The plant used a multi-stage flash distillation process, was twinned with three thermoelectric steam turbine units, and could produce 4.5 mega litres a day. In the end, Burns and Roe ran this facility at Point Loma for less than two years, before the plant became embroiled in political struggles between the US and Cuba. At the end of 1963, as tension between Cuba and the United States escalated, Fidel Castro ordered the water supply to the US military base in Guantanamo Bay to be cut off (Gleick et al., 2009). Faced with this unanticipated water crisis, the US Navy began shipping in potable water at great expense. When relations did not improve, Navy commanders cut the old pipeline, symbolically denying their reliance on Cuba for water. In February 1964 the desalting module at Point Loma was disassembled and shipped to Guantanamo Bay, where it was operated by Burns and Roe for many years.

The Yuma desalination plant also emerged from the intersection between water and international dispute. It was designed and built during the 1970s and 1980s as a political fix for ongoing disagreements between the United States and Mexico over deliveries of Colorado River water (Postel et al., 1998). Construction of the Yuma desalting plant was agreed between President Nixon and President Echeverria in 1974 under the Colorado River Basin Salinity Control Act. For years the United States had been in breach of the quality component of its water delivery commitment to Mexico (Judkins and Larson, 2010). Saline spent irrigation water from the various irrigation districts in the Colorado basin was being returned to the river untreated, meaning that at certain times, the water reaching Mexico had too high a salt content for Mexican irrigators to use. The severity of this problem had become a point of significant political disagreement between the two countries. After lengthy negotiations the US government embarked on a campaign of point-source treatment of agricultural runoff and various irrigation efficiency measures aimed at lowering the salinity of water re-entering the stream and bringing US deliveries of Colorado water to Mexico in line with quality commitments.

The plant at Yuma, located very close to the border, was the flagship infrastructural development of the Salinity Control Act. It was built to desalt agricultural runoff from the Whelton Mohawk Irrigation and Drainage District, which at the time was one of the main polluters in the basin (Taylor and Haugseth, 1976). The design stage of the plant was very long – indeed, although agreed in 1974 the facility did not become operational until 1992. The Bureau of Reclamation used the plant as an opportunity to conduct research and develop desalination technology (Moody et al., 1983). Six different processes were extensively tested at Yuma, both membrane and distillation. After extensive research and development it was decided that spiral-wound reverse osmosis technology was the most promising (Lohman, 1994). So while the facility was built to treat agricultural runoff, its design and operation is very similar to a modern seawater desalter. At the time, its development was cutting edge. Such was the success of the other point-source treatment deployed upstream, however, that since completion in 1992 the Yuma plant has only been operated twice (Bureau of Reclamation, 2015). It stands as an idle monument to the ecological carelessness that has characterised the Bureau of Reclamation's development of the Colorado River.

The scalar fix

The 'desalination fix' has, for the last half-century, emerged periodically under various guises and in various forms (March, 2015). In the 1960s and 1970s saline water conversion twinned with cheap nuclear thermoelectric power was at the heart of optimistic notions of imminent resource abundance – the end of scarcity. The projects proposed to 'fix' the contested politics of the Colorado River during this golden (but ultimately dry) era of desalination were more ambitious in terms of capacity than the projects of today. During the 1980s and 1990s desalination projects of comparably modest size were instead pursued to address specific water governance issues. The combined issues of energy intensity and cost of production prevented desalination from becoming a significant element of municipal water supply for the Colorado River states, and from assuming the position of gateway to the high-technology utopia of its earlier promise (Shiermeier, 2008).

In the contemporary era, the *ocean solution* is once again being pursued as a scalar fix to the insoluble politics of Colorado River water transfers. Desalination has emerged as a powerful discursive and material strategy in paradigmatic shifts, currently underway throughout southwest US and northwest Mexico, towards water supply localisation and diversification. Independent from the contested and inert system of water transfers, desalination is prized as the only "drought-proof supply" that "reduces . . . dependence on water from the Colorado River" (SDCWA, 2017). Desalination is not becoming the catholicon that was once hoped, and indeed, many water agencies have shunned it in favour of more cost-effective diversification strategies. These include the Los Angeles Department of Water and Power, which is pursuing a variety of alternative supplies, including storm water capture, wastewater recycling and groundwater recharge (LADWP, 2015), and Long Beach, which has for a number of years been advancing a highly effective efficiency and conservation agenda (LBWD, 2015). Both of these agencies studied desalination in detail in the 1990s and early 2000s, but deemed it to be too costly.

Nevertheless, desalination has become an important techno-political strategy of water governance for several agencies that have historically relied on Colorado water, most notably San Diego County, which now receives around 10% of its supply from the Carlsbad desalination plant, and Orange County, where a large project in Huntington Beach is under development. Medium-sized projects (with capacities of around 20 mega litres a day) are also being developed across the border in Baja California in La Paz, Cabo San Lucas, Ensenada and San Quintín (D&WR, 2015; McEvoy, 2014). Desalination, in these cases, does not so much represent a silver-bullet solution, but is seen rather to add resilience through diversification of water sources. Thus, cuts to an agency's supply of Colorado water are less damaging if that source represents only one element of a diverse portfolio. This shift towards localisation and diversification, although signalling a broad movement away from the riparian transfer paradigm, does not represent total dis-assembling of its social and technical relations (upon which the Colorado River states still rely), but rather a socio-technical reorientation and assembling of supplementary networks. 'Desalination, the panacea' has become 'desalination, the more or less prudent component of a diversified portfolio.'

The idea for a binational desalination facility emerged from the large water wholesalers on the Colorado River, effectively as a way of 'freeing up' over-allocated water in the Lower Basin and reducing tensions between right-holders. The potential for such a collaboration was first floated by the San Diego County Water Authority (SDCWA) in 2005, who had been studying a number of potential locations for a large facility along the coast on both sides of the border (SDCWA, 2005). A site on Rosarito Beach, around 30 km south of the international border, was identified as particularly favourable, in part because it is adjacent to the Presidente Juarez thermoelectric power plant operated by Comisión Federal de Elecricidad, which would allow the plant developers to take advantage of a number of benefits of infrastructural co-location, including process efficiencies, lower capital costs from shared infrastructure, and simpler permitting from existing industrial land zoning (Pankratz, 2004).

Interest in the Rosarito project increased rapidly and, led by the SDCWA, a total of eight water agencies[1] from both sides of the border embarked on an extensive feasibility and development study (SDCWA, 2010). Two mechanisms for water delivery to the United States were considered: the 'wet water' option and the 'paper water' option. A pipe connecting the desalination plant with San Diego's storage and distribution system could deliver 'wet water' across the border. Alternatively, the entire output of the plant could supply Tijuana and surrounding region, allowing US agencies to withdraw a portion of the output from Mexico's Colorado River allocation. This 'paper transfer' would mean that the desalination plant output would effectively augment the supply available in the Lower Colorado River Basin, allowing the various right-holding agencies to re-allocate water accordingly. For US agencies, the Rosarito project represented a viable way to ease contested Colorado supply. Indeed, the Las Vegas Valley and Southern Nevada had been considering the possibility of financing coastal desalination in return for increased Colorado allocation for some time (Shrestha et al., 2011). On the Mexican side, the Baja state government had attempted a number of times to develop desalination capacity to increase supply reliability. In Tijuana particularly, which relies almost entirely on limited Colorado supply imported from the Mexicali Valley, per-capita water consumption is 30% lower than the Mexican national average and there is widespread belief that inadequate and insecure water supply is retarding the city's growth (Meehan, 2013; Fullerton et al., 2007).

After the completion of the feasibility study, interest from the eight participating agencies slackened somewhat as alternative supply diversification strategies took precedence. For example, the San Diego County Water Authority, which had led the project up to 2010, focussed its attention instead upon securing an alternative transfer deal of Colorado River water with the Imperial Irrigation District, and developing desalting projects in the north of the county (SDCWA, 2013). Nevertheless, the idea for a binational desalting plant at Rosarito was not abandoned. After several phases of restructuring the project is now being delivered through a public-private contract between the State of Baja and a company called NSC Agua, which is owned by the Cayman-based firm, Consolidated Water. Half of the plant's 125,000 mega litre a year capacity has been earmarked for sale to agencies in the United States (Smith, 2016).

Conclusion

The North American desert west is a place of wild water dreams. The monumental efforts undertaken during the twentieth century to deliver water in abundance to an arid and isolated corner of the country were no less extraordinary than the countless schemes that failed. For much of the last hundred years the recurrent dreams of desalting the waters of the Pacific Ocean were firmly in the latter group – always just beyond the horizon of viability. The barriers were generally technical, rather than political, and almost always associated with the dual challenge of cost and energy intensity. Kennedy's dream of abundant water provided by high technology in the atomic age, of desalination as a panacea for scarcity, was never realised. By placing desalination in historical context, this remarkable technological phenomenon is understood, conceptually, as fulfilling a function as a political and scalar fix for the transboundary contestations over allocation on the Colorado River. Desalination, because it is rainfall and climate independent and locally produced, has become a powerful discursive and material tool to address the complex politics that beleaguer terrestrial water, which are insoluble without major political and social change.

The structural tensions that characterise water governance in the arid west stem from the particular historical development of the region. It is from these historical conditions that desalination emerges as a 'solution' that allows thirsty urban economies to secure vital supplies without having to engage with the broader political questions of allocation. Ocean water desalting achieves a double

30 *Joe Williams*

movement in contemporary water governance. First, it essentially provides a way of increasing overall supply in the Colorado Basin, allowing urban regions to access secure supply without entirely disassembling the socio-technical relations and systems that have historically been the catalyst for development in the West. Second, desalination has been enrolled as a powerful discursive and material tool in the broad movement towards localised and diversified water supply portfolios. The extraordinary emergence of desalination for the Colorado River states is, therefore, symptomatic of the historical failures of water governance in the arid west.

Note

1 The eight participating agencies were: (US) San Diego County Water Authority, Central Arizona Water Conservation District, Metropolitan Water District of Southern California and Southern Nevada Water Authority; and (Mexico) Comision Nacional del Aqua, Comision Estatal de Servicios Publicos de Tijuana, Comision Estatal del Agua and Comision Internacional de Limites y Agua Seccion.

Bibliography

Al-Karaghouli, A., Kazmerski, L.L. (2013). Energy consumption and water production cost of conventional and renewable-energy-powered desalination processes. *Renewable and Sustainable Energy Reviews*, 24, 343–356. doi:10.1016/j.rser.2012.12.064

Aviram, R., Katz, D., Shmueli, D. (2014). Desalination as a game-changer in transboundary hydro-politics. *Water Policy*, 16, 609–624. doi:10.2166/wp.2014.106

Bureau of Reclamation. (2015). *Yuma Desalting Plant. Bureau of Reclamation.* Available at www.usbr.gov/lc/yuma/facilities/ydp/yao_ydp.html [accessed 02.01.2016].

Colorado River Board of California. (1944). *Statement on Behalf of California Summarizing Some of the Reasons for Opposition to Proposed Treaty With Mexico Relating to the Colorado River,* 20th March 1944. Glendale, CA.

CRWU [Colorado River Water Users] Committee. (1945). *Mexican Water Treaty. Excerpts From Testimony Before the Senate Foreign Relations Committee in Answer to Ten Arguments Made by Proponents of the Treaty,* 12th March 1945. Colorado River Water Users.

D&WR. (2015). *Mexico Invests Heavily in Six New Desalination Projects.* Available at www.desalination.biz/news/news_story.asp?id=7894 [accessed 04.07.2016].

Department of the Interior. (1963). *Pacific Southwest Water Plan.* Washington, DC: US Department of the Interior.

Ekers, M., Prudham, S. (2015). Towards the socio-ecological fix. *Environment and Planning A*, 47, 2438–2445. doi:10.1177/0308518X15617573

Elimelech, M., Phillip, W.A. (2011). The future of seawater desalination: Energy, technology, and the environment. *Science*, 333, 712–717.

Feitelson, E., Jones, A. (2014). Global diffusion of XL-capacity seawater desalination. *Water Policy*, 16, 1031–1053. doi:10.2166/wp.2014.066

Feitelson, E., Rosenthal, G. (2012). Desalination, space and power: The ramifications of Israel's changing water geography. *Geoforum*, 43, 272–284.

Foster, A., Herlihy, J. (1965). Operating experience at San Diego Flash Distillation Plant. Presented at *The Proceedings of the First International Symposium on Water Desalination, Volume 3.* Washington, DC: US Department of the Interior, Office of Saline Water.

The ocean bountiful? 31

Fritzmann, C., Löwenberg, J., Wintgens, T., and Melin, T. (2007). State-of-the-art of reverse osmosis desalination. *Desalination* 216, 1–3, 1–76, https://doi.org/10.1016/j.desal.2006.12.009.

Fullerton, T.M., Tinajero, R., Cota, J.E.M. (2007). An empirical analysis of Tijuana water consumption. *Atlantic Economic Journal*, 35, 357–369. doi:10.1007/s11293-007-9074-x

Gleick, P., Cooley, H., Cohen, M., Morikawa, M., Morrison, J., Palaniappan, M. (2009). *The World's Water 2008–2009*. Washington, DC: Island Press.

Harvey, D. (1996). *Justice, Nature, and the Geography of Difference*. Cambridge: Blackwell.

Harvey, D. (2006). *The Limits to Capital*. London: Verso.

Harvey, D. (2014). *Seventeen Contradictions and the End of Capitalism*. London: Profile Books.

Holtom, H. and Galstaun, L. (1965). Study of 150 MGD desalted water-power dual plant for Southern California. In *Proceedings of the First International Symposium on Water Desalination*, Volume 3, pp. 543–572. Washington DC: US Department of the Interior, Office of Saline Water.

Judkins, G.L., Larson, K. (2010). The Yuma desalting plant and Cienega de Santa Clara dispute: A case study review of a workgroup process. *Water Policy*, 12, 401–415. doi:10.2166/wp.2009.084

Kamal, I. (2005). Integration of seawater desalination with power generation. *Desalination* 180, 1–3, 217–229, https://doi.org/10.1016/j.desal.2005.02.007.

Kenis, A., Lievens, M. (2014). Searching for 'the political' in environmental politics. *Environmental Politics*, 23, 531–548. doi:10.1080/09644016.2013.870067

Kennedy, J.F. (1961). *Desalination of Water: New Horizons, a Look at New Developments in the Fields of Science and Medicine* [Audio tape]. Washington, DC: National Archives.

Khawaji, A.D., Kutubkhanah, I.K., Wie, J.M. (2008). Advances in seawater desalination technologies. *Desalination*, 221, 47–69. doi:10.1016/j.desal.2007.01.067

LADWP. (2015). *Urban Water Management Plan*. Los Angeles: Los Angeles Department of Water and Power.

LBWD. (2015). *Urban Water Management Plan*. Long Beach: Long Beach Water Department.

Loftus, A., March, H. (2016). Financializing desalination: Rethinking the returns of big infrastructure. *International Journal of Urban and Regional Research*, 40, 46–61. doi:10.1111/1468-2427.12342

Lohman, E. (1994). Proceedings of the IDA and WRPC World Conference on Desalination and Water Treatment Operating report of the largest reverse osmosis desalting plant. *Desalination*, 96, 349–358. doi:10.1016/0011-9164(94)85185-9

March, H. (2015). The politics, geography, and economics of desalination: A critical review. *WIREs Water*, 2, 231–243. doi:10.1002/wat2.1073

March, H., Saurí, D., Rico-Amorós, A.M. (2014). The end of scarcity? Water desalination as the new cornucopia for Mediterranean Spain. *Journal of Hydrology*, 519, 2642–2651. doi:10.1016/j.jhydrol.2014.04.023

McCarran, P. (1945). *Statement to Committee on Foreign Relations, Water Treaty With Mexico*. 79th Congress, 1st Session. Washington DC.

McEvoy, J. (2014). Desalination and water security: The promise and perils of a technological fix to the water crisis in Baja California Sur, Mexico. *Water Alternatives*, 7, 518–541.

Meehan, K. (2013). Disciplining de facto development: Water theft and hydrosocial order in Tijuana. *Environment and Planning D*, 31, 319–336. doi:10.1068/d20610

Moody, C.D., Kaakinen, J.W., Lozier, J.C., Laverty, P.E. (1983). Yuma desalting test facility: Foulant component study. *Desalination*, 47, 239–253. doi:10.1016/0011-9164(83)87078-7

Moore, J. (2015). *Capitalism in the Web of Life*. London: Verso.

Murray Li, T. (2011). Rendering society technical: Government through community and the ethnographic turn at the World Bank in Indonesia. pp. 57–80 in: Mosse, D. (Ed). *Adventures in Aidland: The Anthropology of Professionals in International Development*. New York: Berghahn.

Pahl-Wostl, C. (2015). *Water Governance in the Face of Global Change: From Understanding to Transformation*. New York: Springer.

Pankratz, T. (2004). *Desalination Technology Trends*. Available at www.twdb.texas.gov/publications/reports/numbered_reports/doc/r363/c2.pdf [accessed 13.06.2017].

Peñate, B., García-Rodríguez, L. (2012). Current trends and future prospects in the design of seawater reverse osmosis desalination technology. *Desalination*, 284, 1–8. doi:10.1016/j.desal.2011.09.010

Postel, S.L., Morrison, J.I., Gleick, P.H. (1998). Allocating fresh water to aquatic ecosystems: The case of the Colorado River Delta. *Water International*, 23, 119–125. doi:10.1080/02508069808686757

Reisner, M. (1986). *Cadillac Desert: The American West and Its Disappearing Water*. New York: Viking.

Rieu-Clarke, A., Moynihan, R., Magsig, B. (2015). *Transboundary Water Governance and Climate Change Adaptation: International Law, Policy Guidelines and Best Practice Application*. New York: UNESCO.

Schiermeier, Q. (2008). Water: purification with a pinch of salt. *Nature*, 452, 7185, 260–261.

SDCWA. (2005). *Feasibility Study of Seawater Desalination Development Opportunities for the San Diego/Tijuana Region*. San Diego: San Diego County Water Authority.

SDCWA. (2010). *Rosarito Beach Binational Desalination Plant: Feasibility, Evaluation and Preliminary Design, Phase 1*. San Diego: San Diego County Water Authority.

SDCWA. (2013). *Regional Water Facilities Optimization and Master Plan Update*. San Diego: San Diego County Water Authority.

SDCWA. (2017). *Seawater Desalination*. Available at www.sdcwa.org/seawater-desalination [accessed 23.04.2017].

Seckler, D.W. (1965). *California Water: A Strategy*. Planning and Conservation League Position Papers.

Shrestha, E., Ahmad, S., Johnson, W., Shrestha, P., Batista, J.R. (2011). Carbon footprint of water conveyance versus desalination as alternatives to expand water supply. *Desalination*, 280, 33–43.

Six States Committee. (1944). *In Support of Ratification of the Treaty With Mexico Relating to the Utilization of the Waters of Certain Rivers, Volume 1–3*. Arizona, Colorado, New Mexico, Texas, Utah, Wyoming.

Smith, N. (1984). *Uneven Development: Nature, Capital, and the Production of Space*. Oxford: Basil Blackwell.

Smith, R. (2016). Marketing desalinated seawater from Rosarito Beach to US water users. *MSSC 2016 Annual Salinity Summit*, 27–29 January 2016, Las Vegas.

Swyngedouw, E. (2011). Depoliticized environments: The end of nature, climate change and the post-political condition. *Royal Institute of Philosophy Supplements*, 69, 253–274. doi:10.1017/S1358246111000300

Swyngedouw, E. (2013). Into the sea: Desalination as hydro-social fix in Spain. *Annals of the Association of American Geographers*, 103, 261–270. doi.org/10.1080/00045608.2013.754688.

Swyngedouw, E., Williams, J. (2016). From Spain's hydro-deadlock to the desalination fix. *Water International*, 41, 54–73. doi:10.1080/02508060.2016.1107705

Taylor, I.G., Haugseth, L.A. (1976). Yuma desalting plant design. *Desalination*, 19, 505–523. doi:10.1016/S0011-9164(00)88061-3

Warren, E. (1945). *Mexican Water Treaty: Suggested Revisions*. Statement by Governor Earl Warren before the Foreign Relations Committee of the Senate. (5th February 1945). Washington, DC.

Wilder, M.O., Aguilar-Barajas, I., Pineda-Pablos, N., Varady, R.G., Megdal, S.B., McEvoy, J., Merideth, R., Zúñiga-Terán, A.A., Scott, C.A. (2016). Desalination and water security in the US – Mexico border region: Assessing the social, environmental and political impacts. *Water International*, 1–20. doi:10.1080/02508060.2016.1166416

Wilson, J., Swyngedouw, E. (Eds). (2014). *The Post-Political and Its Discontents: Spaces of Depoliticisation, Spectres of Radical Politics*. Edinburgh: Edinburgh University Press.

3 Piercing the Pyrenees, connecting Catalonia to Europe

The ascendancy and dismissal of the Rhône water transfer project (1994–2016)

Santiago Gorostiza, Hug March and David Saurí

Introduction

In this chapter we will explore the ascendancy and demise of one of the most peculiar water projects of recent decades in Europe: the construction of a 320-kilometre-long pipe to bring water from the Rhône River, in France, to Barcelona.[1] Long-distance water transfers have constituted part of the hydraulic landscape for millennia and many are emblematic of the enormous wealth but also of the terrible social and environmental effects created by moving water. One has only to think of the Colorado River Aqueduct bringing water to Southern California to the detriment of Mexico or the diversions affecting the Amu Darya and Syr Darya rivers in Central Asia that contributed to drying up the Aral Sea to acknowledge the impacts of these hydraulic infrastructures, now dwarfed by the massive mobilization of the Yangtze basin to provide water to the thirsty North China Plain. Unlike these projects, however, the proposal to divert water from the Rhône to the south of the Pyrenees involved two sovereign states and perhaps more significantly was promoted not by these states but by "regions" within them. These two regions, Catalonia on the one hand and Languedoc-Roussillon on the other, could claim a shared historical background. On the other hand, both Catalonia and the Rhône-Alpes region also belonged to the "Four Motors of Europe" (together with Lombardy in Italy and Baden-Württemberg in Germany), a regional lobby in the European Union promoting bilateral agreements on a variety of economic, environmental, cultural, etc. issues for which they are allowed to bypass the national states.

The Rhône transfer embraces many of the topics of interest for a critical view on water planning and management. To begin with, it is another example of the "hydraulic paradigm" (Saurí and Del Moral, 2001) or the historically hegemonic approach to solve the space and time mismatches between water supply and water demand. As Swyngedouw (2015) has argued in the case of Spain, these mismatches are of course "natural" and had to be corrected by human (hydraulic) ingenuity. Likewise, waters from the Rhône would also correct imbalances by connecting the water-rich north with the water-poor south. But the Rhône project had other characteristics that make it a singular example, especially the symbolic meaning for the proponents of the project in Catalonia of a resource that transcended nation-state barriers and strengthened ties with Europe. This chapter uses the contributions of French engineer and planner Bernard Barraqué, consultant on the project from the French side and one of its most incisive critics, to examine the Rhône project as an example of the demise of the "hydro-dinosaurs" (as Barraqué colourfully calls large water transfer schemes) as well as the existence of alternative solutions to the water problems of Barcelona (Rinaudo and Barraqué, 2015; Barraqué et al., 2011; Barraqué, 2004, 2000).

Our main sources have been the press (particularly the main Catalan newspaper, *La Vanguardia*, since 1994), the minutes from the ad hoc expert commission created in Catalonia to study the project (1998–1999), the minutes of several groups opposed to the project located in the historical archive of Girona, and interviews with experts that chaired the technical commission in charge of drafting the different proposals as the project moved along.

Origins: connecting Catalonia to Europe's water tanks (1994–1999)

Looking ahead more than 20 years, to far-off 2015, the Catalan planners of post-Olympics Barcelona saw a calamity looming. According to their forecasts, the significant growth of population predicted for the city and the increasing water consumption trends aimed towards a scenario of water supply crisis. The region not only suffered recurrent droughts, but the bad quality of their supply had already been the cause of substantial investments before the Olympic Games. It was expected that none of the two rivers managed by the Catalan authorities, the Llobregat and the Ter Rivers, which supplied most of the city drinking needs, would suffice in the near future. On the one hand, Llobregat waters were contaminated by several industries and particularly compromised by the activity of potash mining upstream (Gorostiza and Saurí, 2017; Gorostiza et al., 2015; Gorostiza, 2014; Gorostiza et al., 2013). On the other hand, the Ter waters were undoubtedly of a much better quality but came to Barcelona from the north as a result of a water transfer between the Catalan provinces of Girona and Barcelona that had been inaugurated during the Francoist dictatorship, in 1966, and which configured an enduring territorial tension (Jordà-Capdevila, 2016; on the history of Barcelona water supply and consumption, see Tello and Ostos, 2012).

The closest alternative to supplement the city's water supply was in the Ebro River, 200 km to the south. However, the Ebro configured a transregional basin and only the Spanish state could decide its future. The Water Law of 1985 established that transregional river basins (i.e. flowing through at least two regions) fell under the political authority of the Spanish state. While a water transfer towards the north reached the industrial area of Tarragona, it could not be extended to the Catalan capital without the approval of the Spanish government. Moreover, such decision would undoubtedly stir the opposition of the Ebro basin, both inside and outside Catalonia (Jordà-Capdevila and Rodríguez-Labajos, 2014).

Around 1994, and at the same time that the conservative Catalan government demanded the Ebro waters for Barcelona, it started considering an alternative. The Rhône River, in southern France, had a much larger river flow than the Ebro. In comparison, a possible water transfer from the Rhône could guarantee abundant resources with a proportionally minor impact. Despite being farther away (320 km vs. 200 km) and thus more expensive, it also promised water of good quality that would offset the over-salinised flows of the Llobregat and open the door to the restitution of the Ter waters to Girona. Among engineering circles the project became an ideal(ised) alternative to the Ebro which avoided social and territorial conflicts while satisfying Barcelona's water demands. Most of all, from a political standpoint the water transfer would materialise the discursive pro-European stand of the conservative Catalan nationalists into a real physical connection to the European water grid. The water transfer could thus be seen as the materialisation of European integration.

In fact, the new scenario opened by the Spanish accession to the EU in 1986 was fundamental for the initial idea to become a full-fledged project. After some contacts between engineers in 1994 (Desbordes, 2009), the first funds to study the viability of the water transfer came from European regional schemes. Without previous consultation of Spanish authorities, the Catalan government started talks with the French region of Languedoc-Roussillon and later applied for and obtained EU Interreg funds to explore the feasibility of the Rhône water transfer (Espais, 1995; La Vanguardia, 1995). Next year, the Catalan public water company *Aigües Ter-Llobregat* (ATLL) and its French counterpart *Compagnie National d'Aménagement de la Région du Bas-Rhône et du Languedoc* (BRL) joined forces to establish a European Economic Interest Grouping (EEIG) complying with EU regulations (EC 2137–85). Under the name of Languedoc-Roussillon-Catalonia (LRC) aqueduct, a new legal entity came into existence in 1996 (Aigües Ter-Llobregat and Generalitat de Catalunya, 1999).

In the meantime, the Spanish conservatives of *Partido Popular* (PP) had defeated the Spanish socialists and attained the Spanish presidency with the key support of conservative Catalan nationalists (CiU, *Convergència i Unió*). High-level contacts followed between Spanish and French ministries in 1996, and the Spanish foreign minister publically acknowledged the common interest of the project at a joint press conference with his French counterpart. According to the Spanish minister, both the European Commission and the European Investment Bank considered the project as "highly interesting" and would look favourably upon any funding petition (La Vanguardia, 1996). Even the European Parliament adopted a resolution in 1998 "on the technological feasibility of trans-European hydrological networks" which referred explicitly to the Rhône water transfer (Barraqué, 2000; European Parliament, 1998).

In order to assess the environmental impact of the project, two commissions – one in Catalonia, the other in France – started working in 1998. The French committee regarded the impact of the water transfer (quantified in two scenarios of 473 hm^3 per year or 315 hm^3 per year) as negligible for the Rhône river flow in any case. The Catalan committee published its assessment in

1999, underlining that the Rhône River was a better option for a water transfer than the Ebro. According to the views expressed by their report, there were no "scientific objections" to the water transfer project, which would also open the door to European funds and contribute to European integration.[2] In the meantime, the positive atmosphere for the project continued. In 1999, the UNESCO International Hydrological Programme organized a workshop in Paris on interbasin water transfers (UNESCO, 1999). The meeting brought together international experiences on water transfers from South Africa to Australia, but more than two-thirds of the assistants were either Spanish or French. Most of them were scientists, engineers and politicians related to the Rhône water transfer project, which was of course presented at the workshop (Blanc and Imbert, 1999; Vilaró, 1999). The situation seemed mature enough for further developments, but an indispensable step was pending. In order to turn the Rhône water transfer into reality, the French and Spanish states had to establish an international legal treaty.

Building alliances against hydraulic structuralism

When the Rhône water transfer was first proposed, the demographic forecasts and water consumption trends wielded by the Catalan government to justify it were sceptically received by the green movements and part of the political opposition. Already in 1996, both the Catalan pro-independence left (*Esquerra Republicana de Catalunya*, ERC) and the red-green party (*Iniciativa per Catalunya Verds*, ICV) argued that demographic and consumption data were inflated and did not constitute enough cause to put in motion such a massive water transfer. Green movements insisted that with smaller investments in improving the efficiency of water supply networks, together with the promotion of water-saving technologies and water reuse, the Rhône waters would be simply not necessary (Ateneu Barcelonès-Secció Ecologia, 1998; Diari de Sessions del Parlament de Catalunya, 1996).

From a more ideological perspective, another of the critiques used against the project addressed directly its symbolic dimension. Two red-green Catalan MPs declared that conservative Catalan nationalists saw the Rhône waters transfer as an attempt to build a hydraulic sort of "spiritual connection" through the Pyrenees and towards Europe (Vergés, 2002). Along these lines, an international independent committee denounced that the ideological justification of the project could not sustain such a grand public work. According to their views, one of the reasons why the French partners of LRC wanted to send water was "historical solidarity reasons between Languedoc and Catalonia" (Barraqué, 2000). The committee added that the water transfer would make Catalonia closer to France and Europe than the rest of Spain, but stated that what the city of Barcelona needed to endure periodical drought episodes were just "security connections". An investment to improve Llobregat River's water quality, affected by potash mines, would render the Rhône water transfer unnecessary, they argued (Barraqué, 2000).

38   *Santiago Gorostiza, Hug March and David Saurí*

In addition, following the protests of anti-nuclear associations, several groups warned that Rhône waters and sediments may carry traces of radioactive pollution as a result of the nuclear power stations located in its banks. In this regard, the report produced in 1994 by the *Commission de Recherche et d'Information Indépendantes sur la Radioactivité* (CRIIRAD) at the request of Avignon authorities, which found tritium pollution in Rhône groundwaters, was repeatedly referred to throughout these years (CRIIRAD, 2008; Ateneu Barcelonès-Secció Ecologia, 1998; Diari de Sessions del Parlament de Catalunya, 1996).

In the region of Girona, closer to the border, the Defense Group of the River Ter (*Grup de Defensa del Ter*, GDT) was especially active in denouncing the water transfer project, which regarded as an example of old-fashioned policy. The GDT accused the Catalan government of using incorrect data and having other political and economic interests in the project. Apart from pointing to the problem of radiation pollution in waters, they also warned of possible conflicts between farmers with water use rights on both sides of the Pyrenees (GDT, 1999). In fact, farmers in Southern France considered that the project might favour their competitors in Spain (The New York Times, 1999). In late 1999, the GDT promoted public protests against the project and its supporters, such as the Girona Chamber of Commerce, and repeatedly criticised the huge economic and energetic costs of the water transfer (Lloveras et al., 2004). Early in 2000, the GDT was pivotal for the foundation of the Platform Against Water Transfers (*Plataforma d'Oposició als Transvasaments*, POT), which brought together 28 different associations from both sides of the Pyrenees and called for new water-saving and reusing policies.[3] Moreover, it received the political support of the red-green party (ICV), the Catalan pro-independence left (ERC) and, significantly, the Catalan socialists (*Ciutadans pel Canvi*, PSC). CiU Catalan nationalists were almost alone in the political arena in their defence of the project, and their position soon worsened.

"A problem of hardcore Spanish nationalism": the dismissal of the Rhône project

The victory of *Partido Popular* (PP) at the national elections in March 2000 had significant effects on Spanish water politics. After being granted a second term in the Parliament, this time with absolute majority, President José María Aznar pushed ahead the approval of a new version of the Spanish National Hydrological Plan (NHP). The backbone of the NHP was formed by water transfers from Ebro River to the south but also to Barcelona. The projects of the Spanish conservatives were soon contested by a powerful mobilisation in the region close to the mouth of the Ebro River, which gave birth to the Platform in Defence of Ebro River (*Plataforma en Defensa de l'Ebre*). These protests rapidly developed into a wider and heterogeneous movement that called for a "New Water Culture" (*Nueva Cultura del Agua*) and rejected water transfers (Saurí and Del Moral, 2001). In December 2000, in coordination with the Platform Against

Water Transfers, they organised the First Conference for a New Water Culture[4] (*Primeres Jornades per una Nova Cultura de l'Aigua*) (Vilaweb, 2000).

CiU Catalan nationalists continued defending the Rhône project but at the same time entered in negotiations with the Spanish conservatives regarding the NHP. In 2001, both parties joined forces to dismiss the alternative proposals from other parties. Barcelona would receive waters from the Ebro River and funding would be granted to the Catalan government to restore the Ebro Delta (La Vanguardia, 2001a). However, despite the Catalan efforts to include the LRC project as part of the NHP – or at least make a reference to it in the text of the law – the *Partido Popular* showed little interest. In fact, the very same day that the president of BRL – the main partner of the Rhône water transfer in France – presented the project at the environmental commission of the Spanish congress (Diario de Sesiones del Congreso de los Diputados, 2001), José María Aznar called a press conference to settle the matter. Talking at the monastery of Poblet – resting place of the kings of Aragon before unification with the Crown of Castile in 1492 – the Spanish president declared that no water from outside Spanish territory would be mobilised in the NHP. "Spain doesn't have to have water dependencies from anyone; we have enough water resources", he announced. In other words: there was no alternative to Ebro river waters, and the water transfer from Rhône river was out of the question (La Vanguardia, 2001b; El País, 2001).

Without the support of the Spanish government for the project, bringing Rhône waters to Catalonia was impossible. Catalan negotiators with the Spanish conservatives were exasperated by Aznar's words. According to one of the leading members of CiU, the reasons why Aznar rejected the Rhône waters were simply "a problem of hardcore Spanish nationalism, for he considers that it is only Spanish waters that have to provide the backbone of Spain" (La Vanguardia, 2001b, 2001c). Despite the humiliation suffered, CiU ended up supporting the NHP project and in the end obtained a minor return. The final text of the law, approved in 2001, included an article on the possibility of studying alternative waters transfers to the "Spanish hydrologic system" (Boletín Oficial del Estado, 2001).

Clutching at this last straw despite the emphatic negative received, the Catalan government redoubled their efforts to promote the Rhône water transfer during 2002. Closing their ears to the very explicit position of the Spanish conservatives, CiU Catalan nationalists participated in conferences under such telling names as "Water from the Alps for Catalonia"[5] and endorsed reports like "Ebro-Rhône: A viable alternative" (Diari de Sessions del Parlament de Catalunya, 2002). However, the Spanish conservatives' position proved to be set in stone: again and again, they stated that their own studies proved that the Ebro water transfer was much more reasonable (El País, 2002). In the meantime, the social mobilisation against the NHP kept growing and intertwined with the protests against the Rhône project, targeting directly the water transfer model – be it Rhône or Ebro waters – and calling for a change of paradigm in water management. Most of the political opposition to CiU – particularly the greens but also the pro-independence left – sided with social movements and vindicated the principles of the New Water Culture, accusing the Catalan government of protecting private economic interests in the Rhône water transfer.[6]

In 2003 Catalan elections, CiU was ousted from power for the first time since 1980, while the pro-independence left party *Esquerra Republicana de Catalunya* (ERC) saw its representation in the Parliament more than doubled. A triple alliance of socialists (*Partit Socialista de Catalunya*, PSC), ERC and the greens (former communists) established a new government and dismissed the Rhône project. A year later, the Spanish conservatives of *Partido Popular* followed a similar fate to CiU, when they were unexpectedly defeated in the Spanish elections by the Spanish socialist party (*Partido Socialista Obrero Español*, PSOE). The Ebro water transfer, backbone of the NHP, was cancelled soon afterwards. To avoid regional upheaval, however, the PSOE proposed the AGUA plan, which involved the massive deployment of desalination plants in the Mediterranean coast (March et al., 2014).

Rhône's waters in Barcelona's pipes: the 2007–2008 drought

Few could have imagined in 2004 that the Rhône waters would arrive anytime soon to Barcelona. After the rejection of the Ebro transfer, a desalination plant with a capacity of 60 cubic hectometres per year was to be built in El Prat de Llobregat, near Barcelona. Its operation was supposed to shelve the debate on the need of transferring Ebro or Rhône waters. However, by 2007 the plant was still not functioning – a circumstance which proved to be critical, because during the years 2007–2008 the worst drought in 50 years struck the metropolitan area of Barcelona.

As the water reserves diminished during 2007, the Catalan government issued several orders to restrict water usage, particularly private and public outdoor water uses (March et al., 2013). However, by the end of the year the drought continued and the reserves of the Ter-Llobregat system providing water to Barcelona were alarmingly low. As the Catalan and Spanish governments – by then both led by the socialist party – searched for solutions such as water shipping to Barcelona in order to confront the coming scenario of water cuts, the Rhône water transfer made a strong comeback. Many of the original proponents of the project, by then in the opposition, saw the coming crisis as the proof that the Rhône water transfer was needed, and felt their former efforts vindicated by the circumstances. According to CiU, the French waters were the only "definitive solution" to water scarcity in Barcelona, and the party vowed to take the project to the Spanish parliament once again (El Mundo, 2008). Soon afterwards, the Catalan Association of Engineers (*Col·legi d'Enginyers de Catalunya*) published a report endorsing the possibility of the Rhône water transfer (Col·legi d'Enginyers de Catalunya, 2008).

By the spring of 2008, the situation of water supply in the region of Barcelona had become dramatic. As in medieval times, the Archbishop of Barcelona made a public call for pro-pluvia rogations. The ecosocialist Catalan Minister admitted

Piercing the Pyrenees 41

that, despite being agnostic, he had asked the Virgin of Montserrat to intercede (La Vanguardia, 2008a). On a perhaps more realistic approach, the Catalan government had also paid enormous sums of money to get water shipped from Tarragona but also, ironically, from Marseille, in the mouth of the Rhône River. Three tankers and four trains loaded with water were the first shipped to sustain the city's water supply during the summer (La Vanguardia, 2008b). After all, the Rhône waters were finally to arrive to Barcelona, not piped but shipped. The irony did not pass unnoticed to the former proponents of the water transfer and their related opinion makers, who wielded the project against the progressive alliance in power, agitating the old doubts about the water quality and accusing the government of lack of foresight (Col·legi d'Enginyers de Catalunya, 2008; La Vanguardia, 2008c, 2008d).

However, the coming crisis was an urgent matter and in any case the Rhône water transfer would take years. Other structural solutions were formulated, and under a situation described as a "national emergency", the water transfer between Tarragona and Barcelona that would take the Ebro waters to the Catalan capital was approved by the Spanish government. Despite the disagreements with its Catalan counterparts, the Socialist government in Madrid also agreed to study the Rhône alternative (March and Saurí, 2013). In May 2008, the first ships loaded with water arrived in Barcelona as the work of the Ebro transfer developed as quickly as possible. At the same time, newspapers reported on contacts at the higher level between Spain and France regarding the Rhône water transfer (La Vanguardia, 2008e). On May 21, 2008 the first tanker loaded with Rhône waters entered the Barcelona harbour, but very few followed (La Vanguardia, 2008f). Rains finally started pouring in mid-May and the Catalan reservoirs swelled fast, rendering water shipping unnecessary. The Ebro connection was also cancelled soon afterwards and the drought declared over in early 2009. Months later, when the Spanish government published its assessment of the Rhône water transfer, it highlighted that it was "expensive, unnecessary and with a high environmental impact". In the words of the Spanish delegate, it was better to support an option in the Spanish national territory that "would not be exposed to the decision of another country" – in reference to France (La Vanguardia, 2009). In July the same year, the desalination plant of El Prat del Llobregat finally started its production of water from the sea.

Reanimating the "water dinosaurs" (2011–2015)

When the dust of the drought crisis in Barcelona had finally settled and the desalination plant was already working, several powerful associations made a renewed vow for the Rhône project. In late 2009, the Catalan Employers Association, the Barcelona Chamber of Commerce, the Association of Engineers and several other entities stated once again that the Rhône waters, "thinking beyond 2020", were badly needed (Fulls dels Enginyers, 2009).

The promotion of the Rhône project remained explicitly mentioned in the electoral platform of the conservative Catalan nationalists, who came back to power in 2010 after defeating the progressive coalition in power (Convergència i Unió, 2010:133). Next year, the Spanish conservative *Partido Popular* also returned to power and left the Socialists in the opposition. However, with the desalination plants completed and actually underused, reasons for the Rhône water transfer seemed even shakier than before.

It did not take long until the by-then Catalan president Artur Mas, who had launched the project back in 1995 as Catalan Minister for Public Works, alluded indirectly to the possibility of establishing "stable connections" with European partners with abundant resources (Nació Digital, 2011a). The Catalan Minister of Territory and Sustainability argued that the Rhône water transfer could not be discarded because the Catalan water regime was very irregular and climate change might make it even more unstable (Nació Digital, 2011b). A timely report published by the Barcelona Chamber of Commerce supplied some new arguments along with the usual repertory. Under the dull title "Hydric resources in Catalonia. Basic Concepts and Data", the report included in its last section an unabashed defence of the project (Dolz and Armengol, 2011:187). Even though many years passed since the late 1990s, its authors considered that the European Union would probably give economic and political support to the water transfer project. Regarding the position of the Spanish government, they acknowledged that such a project would establish "an important bond" between Catalonia and France and thus might be rejected. They suggested that this circumstance might be avoided if the water transfer would also benefit the water supply of the region of València, south of Catalonia (Dolz and Armengol, 2011:187).

Maybe more interestingly, according to the authors of the report, there was a novelty in the French side of the Pyrenees that had created a different scenario in comparison to the 1990s. Under the name of "Aqua Domitia", the regional government of Languedoc had launched a project to transfer water from the Rhône to several cities and tourist resorts in the southeast, thus bringing the much-desired "connection to the Alps water tanks" nearer to Catalonia. The authors emphasised that this new scenario was an opportunity for the Catalan authorities because the costs of bringing water to Barcelona would decrease and new synergies were possible (Dolz and Armengol, 2011:165–166, 170, 173).

A few months later, in February 2012, the Catalan and Languedoc presidents gave a joint press conference after a meeting of the Pyrenees-Mediterranean Euroregion, and Artur Mas announced that he would request the water transfer to the Spanish government, with the support of the Languedoc region (El País, 2012). On another occasion, the pro-European discourse of CiU Catalan conservatives was intertwined with the project to bring French waters to Catalonia. However, the Aqua Domitia project represented only 10% of the original water transfer studied in the late 1990s, and therefore it could not simply be extended to Barcelona. Moreover, it faced its own difficulties and opposition in southern France (Rinaudo and Barraqué, 2015; La Vanguardia, 2012). Although statements regarding the resurrection of the Rhône water transfer coming from the Spanish conservatives followed the declaration of the Catalan president, little on the real progress of the initiative – if any existed – was made public (Diari de Sessions del Parlament de Catalunya, 2014; La Vanguardia, 2014; El País, 2013).

After 2010, the full emergence of the pro-independence movement eroded the relations between the Spanish and Catalan governments. In addition, the 2012 snap elections in Catalonia boosted the pro-independence left (ERC), with a more straight-ahead vision on Catalan secession from Spain. ERC rejected inter-basin water transfers and supported the principles of the New Water Culture. After the breaking of the long-lasting coalition of Catalan conservatives (*Convergència i Unió* – CiU) and yet another round of snap elections in 2015, a new alliance was formed. The liberal *Convergència Democràtica de Catalunya* – originally part of CiU – joined ERC and the main civil associations promoting independence to establish *Junts pel Sí* ("Together for Yes"), a candidacy almost solely focused on the independence debate. Their electoral platform proposed a future "national pact" on water based on the principles of the New Water Culture and made no mention of water transfers (Junts pel Sí, 2015:104–105). At these last elections only the former allies of Convergència – *Unió Democràtica de Catalunya* – included the Rhône project in its electoral platform, and failed to win any seats in the Parliament.

Conclusions

At the moment, within the emergent approaches of efficiency and alternative resources (desalination, wastewater reuse, etc.), as well as a context of declining water consumption, the connection of the Catalan capital to "Europe's water tanks" seems highly unlikely. Interestingly, during the last years the project has arguably not occupied a central place in the popular imagery of surging Catalan secessionism, despite the repeated critiques against Spanish political centralism regarding infrastructure. Nonetheless, it has percolated into some of the legal documents drafting the features of a hypothetical Catalan state. According to the white paper on the National Transition of Catalonia, for instance, the risk of failure of the Catalan water supply system has been "greatly reduced", but the "Rhône transfer" should be studied as an alternative water supply measure in the long term (Generalitat de Catalunya, 2014:96–98).

Since the mid-1990s, the Rhône water transfer has been the flagship water project of CiU Catalan conservatives. Taking advantage of the new scenario created by the Spanish admittance to the EU, it posited a national reconfiguration through water infrastructures. Catalan conservative nationalists have obstinately stuck to the project throughout the years – with the support of the employers' associations, chambers of commerce, associations of engineers and many others – even after the approval of the EU Water Framework Directive or the massive Spanish and EU investments in the AGUA desalination plan. The Rhône waters have been imagined and presented as the panacea that would alleviate all of Catalonia's internal water conflicts, easing the pressure on the Ter and Ebro rivers. Despite these attempts to draw popular support, the movements against water transfers that emerged around 2000 successfully mobilised a heterogeneous group of citizens and scientists whose demands challenge the project's justification. Moreover, territorial alliances between movements protecting different rivers mostly circumvented possible tensions between themselves, while their joint critiques to the supply approaches championed by the water transfer proponents hit below the waterline of hydraulic structuralism. Lastly, without the support of the rest of the Catalan political spectrum – particularly from the growing pro-independence left – Catalan conservative nationalists failed to build a wide consensus for the water transfer.

The Rhône water transfer proponents also failed to overcome the ingrained national(istic) principle of Spanish hydraulic policy, as shown by the emphatic refusal of the Spanish conservatives to explore this alternative in 2001 while at the same time advocating for the (Spanish) National Hydrological Plan. Despite the shiny pro-European discourse of the Catalan conservatives, it goes without saying that the connection of Catalonia to the European water grid was no less touched by the "hydraulic structuralism" doctrine than the homogenising Spanish NHP (Sauri and Del Moral, 2001). The Rhône water transfer project was based on the very same creed, but it attempted to produce new scales of hydro-social cycle to guarantee waters and satisfy Catalan economic elites and water lobbies. Although the project has been consistently rejected by the Catalan pro-independence left, it keeps resurfacing time and again. In December 2015, in a conference about water supply at the Girona chamber of commerce, an engineering professor and author of a report on water supply in Catalonia censured the pro-independence left for not supporting the Rhône project and stated that only a connection to "the big European water tank" would mean "almost the final solution" for Catalonia's water supply (Ara, 2015). More and more, recent formulations of a hypothetical Rhône water transfer project look beyond 2020 and refer to climate change as an additional reason to secure water supply in the region (El Punt Avui, 2012; Dolz and Armengol, 2011). The Rhône water transfer may well be, as put by Bernard Barraqué, a "dying hydro-dinosaur" of a past age, but it is proving to die hard.

Acknowledgments

Santiago Gorostiza acknowledges the financial support from the Spanish Ministry of Economy and Competitiveness, through the "María de Maeztu" program for Units of Excellence (MDM-2015-0552).

Notes

1 This chapter develops some of the arguments presented in Gorostiza et al. (2017), adding new data and information.
2 "Informe del Comitè Científic Assessor sobre la situació de l'abastament d'aigua a l'àrea de Barcelona", April 29, 1999. We thank Robert Vergés for granting us access to the minutes of the committee's meetings.
3 Diari de Girona, January 14, 2000, "Girona crea una plataforma contra els transvasaments" and minutes of the assembly of the Platform of Opposition to Water Transfers, April 11,

2000, both at the Arxiu Històric de Girona (AHG), Associació Naturalista de Girona (ANG), 81–554. We are grateful to Dídac Jordà-Capdevila for providing information regarding this matter.

4 Minutes of the assembly of the Platform of Opposition to Water Transfers, November 4, 2000, AHG, ANG, 81–554.

5 November 19, 2002, Jornada "Aigua dels Alps per a Catalunya", *RiverNet*, www.rivernet. org/rhonebarcelone/eaudesalpes_cat.htm. Last accessed April 29, 2017.

6 "Vendrell afirma que la proposta de transvasament del Roine respon a interessos econòmics particulars", May 24, 2002, www.esquerra.cat/actualitat/vendrell-afirma-que-la-proposta-de-transvasament-del-roine-respon-a-interessos-economics-p#sthash.glSNDHBT.dpuf. Last accessed September 10, 2016.

References

Aigües Ter-Llobregat and Generalitat de Catalunya. (1999). *L'Abastament d'aigua a les comarques de l'entorn de Barcelona*. Barcelona, Spain: Generalitat de Catalunya and Aigües-Ter Llobregat.

Ara. (2015, December 4). *Josep Dolz (UPC) defensa el transvasament del Roine i la interconnexió hídrica com a solució per al dèficit hídric*. Retrieved from: www.ara.cat/comarquesgironines/Josep-Dolz-UPC-Roine-interconnexio_0_1479452172.html. Last accessed May 2, 2017.

Ateneu Barcelonès-Secció Ecologia. (1998). ¿És recomanable el transvasament del Roine?. *Conference at Ateneu Barcelonès*, November 30, 1998. Recording available at the collection of Ateneu Barcelonès, Barcelona.

Barraqué, B. (2000). *Executive Summary LRC Project*. Retrieved from: www.hydrologie.org/hydrodinosaures/Rhône_ang.htm. Last accessed May 1, 2017.

Barraqué, B. (2004). The Three Ages of Engineering for the Water Industry. *Anuari de la Societat Catalana d'Economia*, 18, 135–152.

Barraqué, B., Juuti, P.S., Katko, T.S. (2011). Urban Water Conflicts in Recent European History: Changing Interactions Between Technology, Environment and Society. In Barraqué, B. (Ed.), *Urban Water Conflicts*. Paris and Leiden: UNESCO Publishing & CRC Press, pp. 15–32.

Blanc, J.L., Imbert, F. (1999). L'aqueduct Languedoc-Roussillon-Catalogne. In UNESCO (Ed.), *Proceedings of the International Workshop on Interbasin Water Transfer*, 25–27 April 1999. Paris, France: UNESCO, pp. 33–44.

Boletín Oficial del Estado. (2001). Ley 10/2001, de 5 de julio, del Plan Hidrológico Nacional. *BOE*, 161, 24228–24250.

Col·legi d'Enginyers de Catalunya. (2008). *L'aigua: un fre al desenvolupament*. Retrieved from: www.eic.cat/2008-laigua-fre-desenvolupament. Last accessed May 1, 2017.

Convergència i Unió. (2010). *Eleccions Nacionals 2010. Programa de govern, projecte de país*. Barcelona: Convergència i Unió.

CRIIRAD. (2008, Octobre). Marcoule et tritium. *Note CRIIRAD N°08–159*. Retrieved from: www.criirad.org/installations-nucl/marcoule_tritium.pdf. Last accessed April 30, 2017.

Desbordes, M. (2009). Une histoire d'aqueduc: Montpellier-Barcelone. *Revue de l'économie méridionale*, 57(227), 199–210.

Diario de Sesiones del Congreso de los Diputados. (2001). Comisión de Medio Ambiente. "Del presidente del Directorio del Consorcio BRL, organismo gestor de la explotación de la concesión del río Ródano (M. Jean Pierre Brunel)". *Número de expediente 219/000169.*

Diari de Sessions del Parlament de Catalunya. (1996). Sessió informativa del Conseller de Política Territorial i Obres Públiques perquè informi sobre el transvasament d'aigües del riu Roine. *Diari de Sessions del Parlament de Catalunya*, C(65), 1807–1829.

Diari de Sessions del Parlament de Catalunya. (2002). Sessió informativa amb el conseller de Medi Ambient per informar sobre el contingut de l'estudi "Ebre – Roine: una alternativa viable". *Diari de Sessions del Parlament de Catalunya*, C(414), 5–59.

Diari de Sessions del Parlament de Catalunya. (2014). Interpel·lació al Govern sobre el Pla Hidrològic de la conca de l'Ebre i sobre el Pla de protecció integral del delta de l'Ebre (tram. 300–00157/10). *Diari de Sessions del Parlament de Catalunya*, P(56), 86–91.

Dolz, J., Armengol, J. (2011). *Els recursos hídrics a Catalunya. Dades i conceptes bàsics.* Barcelona: Cambra de Comerç, Indústria i Navegació de Barcelona. Retrieved from: www.upc. edu/saladepremsa/al-dia/mes-noticies/presentat-un-informe-sobre-els-recursos-hidrics-a-catalunya/Informe-Els-recursos-hidrics-a-Catalunya.pdf. Last accessed May 1, 2017.

El Mundo. (2008, January 14). *Duran llevará al Congreso el proyecto para trasvasar agua del río Ródano.* Retrieved from: www.elmundo.es/elmundo/2008/01/13/barcelona/1200252131. html. Last accessed April 30, 2017.

El País. (2001, March 28). *Aznar niega a Pujol el trasvase de agua del Ródano para Cataluña.* Retrieved from: http://elpais.com/diario/2001/03/28/espana/985730409_850215.html. Last accessed April 30, 2017.

El País. (2002, July 22). *El trasvase del Ródano costará el doble que el del Ebro, según el Gobierno central.* Retrieved from: http://elpais.com/diario/2002/07/22/catalunya/1027300058_850215. html. Last accessed April 30, 2017.

El País. (2012, February 13). *Mas planteará a Rajoy el trasvase del Ródano como alternativa al del Ebro.* Retrieved from: http://ccaa.elpais.com/ccaa/2012/02/13/catalunya/1329150493_ 622173.html. Last accessed May 1, 2017.

El País. (2013, November 5). *Medio ambiente recupera el trasvase del Ródano.* Retrieved from: http://sociedad.elpais.com/sociedad/2013/11/05/actualidad/1383683735_763655.html Last accessed May 1, 2017.

El Punt Avui. (2012, February 27). *L'equació sense el Ter.* Retrieved from: www.elpuntavui. cat/article/1-territori/11-mediambient/510686-lequacio-sense-el-ter.html. Last accessed May 2, 2017.

Espais. (1995, July). Revista del Departament de Política Territorial i Obres Públiques. *Acord per estudiar el transvasament d'aigua del Roine*, p. 6.

European Parliament. (1998, January 29). *A Trans-European Water Network (A4–407/97 – Izquierdo Collado).* Retrieved from: www.europarl.europa.eu/press/sdp/pointses/en/1998/ p980128.htm#15. Last accessed May 3, 2016.

Fulls dels Enginyers. (2009, November–Desember). *La patronal aprofita el dia per reclamar els transvasaments de l'Ebre i del Roine no previstos*, n°264, p. 45.

Generalitat de Catalunya. (2014). *White Paper on The National Transition of Catalonia. Synthesis.* Barcelona: Generalitat de Catalunya.

Gorostiza, S. (2014). Potash Extraction and Historical Environmental Conflict in the Bages Region (Spain). *Investigaciones Geográficas*, 61, 5–16.

Gorostiza, S., Honey-Rosés, J., Lloret, R. (2015). *Rius de Sal: una visió històrica de la salinització dels rius Llobregat i Cardener durant el segle XX.* Sant Feliu de Llobregat: Edicions del Llobregat, Centre d'Estudis Comarcals del Baix Llobregat.

Gorostiza, S., March, H., Sauri, D. (2013). Servicing Customers in Revolutionary Times: The Experience of the Collectivized Barcelona Water Company During the Spanish Civil War. *Antipode*, 45, 908–925.

Gorostiza, S., Saurí, D. (2017). Dangerous Assemblages: Salts, Trihalomethanes and Endocrine Disruptors in the Water Palimpsest of the Llobregat River, Catalonia. *Geoforum*, 81, 153–162.

Gorostiza, S., March, H., Saurí, D. (2017). Flows from beyond the Pyrenees. The Rhône River and Catalonia's search for water independence. *Political Geography*, 60, 132–142.

Grup de Defensa del Ter. (1999, February 4). *Statement of the Grup de Defensa del Ter*. GDT.

Jordà-Capdevila, D. (2016). *Water Flows to Multiple Stakeholders: An Ecosystem Services-based Approach to Conflicts in the Ter River Basin*. PhD dissertation, Universitat Autònoma de Barcelona.

Jordà-Capdevila, D., Rodríguez-Labajos, B. (2014). An Ecosystem Service Approach to Understand Conflicts on River Flows: Local Views on the Ter River (Catalonia). *Sustainaibility Science*, 10, 463–477.

Junts pel Sí. (2015). *Programa electoral*. Retrieved from: http://juntspelsi.s3.amazonaws.com/assets/150905_Programa_electoral_v1.pdf. Last accessed May 1, 2017.

La Vanguardia. (1995, February 9). La Generalitat negocia con Languedoc traer agua del Ródano. *La Vanguardia*, p. 1.

La Vanguardia. (1996, July 19). Francia y España estudiarán el trasvase de aguas del Ródano. *La Vanguardia*, p. 17.

La Vanguardia. (2001a, March 23). CiU vota con el PP contra la retirada del plan hidrológico. *La Vanguardia*, p. 1.

La Vanguardia. (2001b, March 28). Aznar advierte que no hay alternativa al trasvase del Ebro. *La Vanguardia*, p. 1, 13.

La Vanguardia (2001c, March 30). Trias atribuye a un "nacionalismo rancio" la negativa de Aznar al trasvase del Ródano. *La Vanguardia*, p. 16.

La Vanguardia (2008a, April 3). Sistach llama a rezar para pedir la lluvia. *La Vanguardia*. Retrieved from: www.lavanguardia.com/vida/20080403/53452201611/sistach-llama-a-rezar-para-pedir-la-lluvia.html. Last accessed May 1, 2017.

La Vanguardia. (2008b, April 4). Zapatero rechaza con total contundencia la opción del Segre. *La Vanguardia*, p. 1.

La Vanguardia. (2008c, April 4). Agua: imprevisión y engaños. *La Vanguardia*, p. 25.

La Vanguardia. (2008d, April 4). Ahogados sin agua. *La Vanguardia*, p. 28.

La Vanguardia. (2008e, April 15). Zapatero ha hablado del Ródano con Sarkozy. *La Vanguardia*, p. 1.

La Vanguardia. (2008f, May 21). Llega el agua del Ródano. *Vivir en Barcelona, La Vanguardia*, p. 4.

La Vanguardia. (2009, May 6). Varapalo al trasvase del Ródano. *La Vanguardia*. Retrieved from: www.lavanguardia.com/vida/20090506/53696972651/varapalo-al-trasvase-del-rodano.html. Last accessed May 1, 2017.

La Vanguardia. (2012, July 10). França no espera i llança un minitransvasament del Roine. *Viure a Barcelona, La Vanguardia*, p. 7.

La Vanguardia. (2014, April 20). La reaparición del Ródano. *La Vanguardia*, p. 17.

Lloveras, M., Sivillà, I., Valls, P. (2004, Novembre). La Plataforma d'Oposició als Transvasaments. *Suport a la gestió ambiental d'activitats en el municipi*, pp. 134–141.

March, H., Domènech, L., Saurí, D. (2013). Water Conservation Campaigns and Citizen Perceptions: The Drought of 2007–2008 in the Metropolitan Area of Barcelona. *Natural Hazards*, 65, 1951–1966.

March, H., Saurí, D. (2013). La sequera del 2007–2008 a la Ciutat de Barcelona: gènesi, gestió i visions discordants. *Treballs de la Societat Catalana de Geografia*, 76, 289–306.

March, H., Saurí, D., Rico-Amorós, A.M. (2014). The End of Scarcity? Water Desalination as the New Cornucopia for Mediterranean Spain. *Journal of Hydrology*, 519, 2642–2651.

Nació Digital. (2011a, April 1). *Mas revifa el fantasma del Roine*. Retrieved from: www.naciodigital.cat/noticia/23727/mas/revifa/fantasma/roine. Last accessed May 1, 2017.

Nació Digital. (2011b, July 1). *El Roine a la memòria*. Retrieved from: www.naciodigital.cat/noticia/33116/roine/memoria?rlc=a2. Last accessed May 1, 2017.

The New York Times. (1999, July 19). Water Scarce, Barcelona Plans Big Pipe to Tap Rhône. *The New York Times*. Retrieved from: www.nytimes.com/1999/07/19/world/water-scarce-barcelona-plans-big-pipe-to-tap-Rhône.html. Last accessed April 30, 2017.

Rinaudo, J.-D., Barraqué, B. (2015). Inter-Basin Transfers as a Supply Option: The End of an Era? In Grafton, Q., Daniell, K., Nauges, C., Rinaudo, J.-D., Chan, N. (Eds.), *Understanding and Managing Urban Water in Transition.* Dordrecht: Springer Netherlands, pp. 175–200.

Saurí, D., Del Moral, L. (2001). Recent Developments in Spanish Water Policy. Alternatives and Conflicts at the End of the Hydraulic Age. *Geoforum*, 32(3), 351–362.

Swyngedouw, E. (2015). *Liquid Power: Contested Hydro-Modernities in Twentieth-Century Spain.* Cambridge, MA and London: The MIT Press.

Tello, E., Ostos, J. (2012). Water Consumption in Barcelona and Its Regional Environmental Imprint: A Long-term History (1717–2008). *Regional Environmental Change*, 12(2), 347–361.

UNESCO. (1999). *Proceedings of the International Workshop on Interbasin Water Transfer*, Paris, 25–27 April 1999. Paris: UNESCO.

Vergés, J.C. (2002). *El saqueo del agua en España.* Barcelona: Ediciones de la Tempestad.

Vilaró, F. (1999). Abastecimiento a la región de Barcelona. In UNESCO (Ed.), *Proceedings of the International Workshop on Interbasin Water Transfer*, Paris, 25–27 April 1999. Paris: UNESCO, pp. 17–31.

Vilaweb. (2000, December 17). *S'inauguren les 'Jornades catalanes per una nova cultura de l'aigua'.* Retrieved from: www.vilaweb.cat/noticia/1158536/20001217/sinauguren-jornades-catalanes-nova-cultura-laigua.html. Last accessed April 30, 2017.

4 Death by certainty

The Vinça dam, the French state,
and the changing social relations of
the irrigation of the Têt basin of the
Eastern French Pyrénées

Jamie Linton and Etienne Delay

Water, technology and the nation state

To begin by discussing the trilogy of terms that frame this volume, we view water as a relational outcome, or hybrid, of socionature (Swyngedouw, 2004; 1999). Water therefore becomes something that differs in relation to the circumstances – epistemological, technological, infrastructural, political, economic, etc. – in which it gets produced. The waters that come into play in the present study can be described first, in terms of "modern water", which we have elsewhere described as a mode of knowing and representing water that has dominated modern hydrological discourse, especially in the more industrialized parts of the world. This modern way of knowing and representing water essentially abstracts all waters from the social, historical, and local conditions in which they are produced and reduces them to a common abstract and timeless identity; it is conducive to a style of hydro-social relations that is reflected in the idea of 'water resources' and the practice of 'water management' (Linton, 2014; 2010). Modern water was produced by the French state through the process of impoundment and release at a dam built in 1976 for the twin purposes of assuring downstream flows in the summer season and regulating flood runoff in the fall season. This kind of water and style of management may be contrasted with the water flowing in irrigation canals below the dam and distributed to farmers' crops by means of traditional gravity field irrigation techniques. Although this water is partly regulated by the dam, it flows according to the logic of more ancient rules of collective management and distribution that evolved since the Middle Ages and structure networks of social relations that are based on the principles of mutual aid and produce water as a shared resource. A third type of water occurs by means of pumps, pipes, and irrigation (spray and drip) technologies that stand in contrast to the gravity-based irrigation technologies of the region. These latter technologies structure transactional social relations corresponding more closely to modern water, and produce water as a commodity.

Technology, in our view, is equally a relational outcome of these same processes, just as it contributes to the production process of other components of what has been called the hydro-social cycle (Linton and Budds, 2014). Obertreis et al. (2016) have put it nicely for infrastructure, which we consider an aspect of technology:

> Building on insight from science and technology studies, social scientists and historians are in wide agreement that an infrastructure system for, say, water irrigation or supply cannot be reduced to its material/physical components alone. Instead, it needs to be seen as a combination of technical artifacts, regulatory frameworks, cultural norms, environmental flows, funding mechanisms, governance forms, etc. that get configured in particular ways in particular places at particular times. The significance of this socio-technical understanding is not simply that infrastructure systems are more complex than previously conceived, but that they co-evolve in myriad relations between society, nature and technology. This relational understanding of infrastructure as being part of broader societal and environmental structures and processes but also itself consisting of social and ecological dimensions has opened up new avenues for understanding the societal constitution and workings of infrastructure.
>
> (Obertreis et al., 2016: 172)

The nation-state is less a thing in itself than a network of heterogeneous actors that might be considered a relational accomplishment, something that is continually affecting socionature and changing in relation to changes in socionature (Alex Loftus, pers.com, October 2016). The main agency, or structure, through which the French state is involved in the region, is the Department of Pyrénées Orientales (hereafter referred to as Eastern Pyrénées). The state is also present in the form of the Agence de l'Eau Rhône Méditerranée Corse (the regional Water Agency), which develops regulations in order that the requirements of French environmental law and the European Water Framework are met in the region. It is the department that was responsible for constructing the dam at Vinça and which governs its management. Between the time when the dam was built in the 1970s and the present, the Department has evolved from an authority that produces modern water for (modern) agricultural production and flood protection to one that strives to produce multiple waters in an era of environmental governance. The Water Agency offers financial incentives to promote water-use efficiency and other practices that treat, and produce, water as a scarce resource and commodity as well as an ecological entity. Through these instruments, the Water Agency and the Department promote pressurized (spray and drip) irrigation as the main solution to perceived water problems.

We are interested in the implications of the Vinça dam – and other infrastructures such as irrigation canals and water pumps – for the social relations that constitute irrigation in the lower Têt basin. This interest leads us to consider the relevance of Wittfogel's hydraulic society thesis and with the water-society dialectic that it has helped inspire (Wittfogel, 1957). Wittfogel's work on the history of hydraulic society has been influential in the study of hydro-social relations, especially incorporating technology and infrastructure (see for example Bichsel, 2016; Obertreis et al., 2016; Banister, 2014). While eschewing the (environmentally) deterministic aspects of Wittfogel's work, we adopt the basic dialectical insight that the ways people manage water and deal with water problems can have an important influence on social relations, which in turn may affect hydrological processes (Worster, 1985: 22). We take Wittfogel's basic insight that the way people respond to the need to manage water (especially but not exclusively) in arid environments has an important influence on the evolution of social relations – and on the structure of society itself.

Wittfogel's study of ancient hydraulic societies led him to conclude that controlling and managing water for irrigation is necessarily a collective effort, which gave rise to – and had the effect of strengthening the authority of – a central state, eventually producing despotic regimes that subjugated individual agricultural producers. Elinor Ostrom (1992) among others challenged this vision showing that while irrigation indeed necessitated a collective effort, groups of farmers themselves could, and historically often did, build and manage irrigation systems on their own, without the involvement of a powerful central state (Ruf, 2011; Ostrom, 1992). Research has long shown that the infrastructure and rules of irrigation in this region were developed and maintained by local associations of farmers grouped in canal associations (Ruf, 1998; de Passa, 1821). As such, the history of irrigation in the basin corresponds more closely to Ostrom's model of participatory irrigation than to Wittfogel's hydraulic society. However as we will show, the role of the state, while largely absent until the mid-twentieth century has grown in tandem with the construction and operation of the dam. This recent growth in the presence and authority of the state derives partly from the obvious need for state financing and expertise in the construction and operation of a large dam and supporting infrastructures. But we argue that the indirect and less obvious effect of the dam in producing what might be called hydrological certainty has been most influential in opening a space for the state to assume control of water management while simultaneously disintegrating the local social networks and relations that had evolved as a response to conditions of hydrological uncertainty.

Geographical and historical context

The geographical context for this research is Têt River basin in the Eastern Pyrénées Department of southern France, which is on the border with Spain. The larger region, known as the Roussillon, is characterized by a Mediterranean climate, featuring impetuous rivers that alternate between periods of severe drought and episodes of devastating floods. Aridity is the general condition during the summertime, meaning that irrigation is required for agriculture, with the general exception of vineyards in the lower basin.

Canals drawn from the Têt River date from the Middle Ages, when the region was part of Catalonia, and originally served mainly industrial (milling and textile production) purposes. These canals were owned, managed, and maintained collectively by property holders (seigneuries, abbeys) granted feudal rights by the

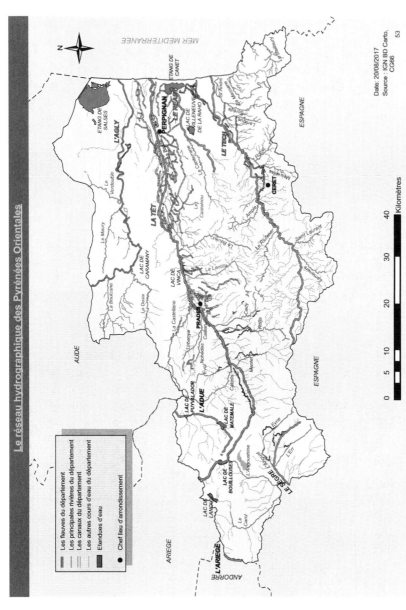

Figure 4.1 Map of the Department of Eastern Pyrénées, southeastern France, with the Têt River shown as the principal river of the Department. The reservoir formed by the Vinça Dam (Lac de Vinça) is visible in the centre. The primary irrigation canals are evident downstream of the dam.

(Source: Department Pyrénées Orientales www.ledepartement66.fr/390-les-rivieres.htm)

Death by certainty 53

Count of Barcelona, who owed fealty to the Crown of Aragon (Ladki et al., 2012: 28). Irrigation grew in importance and became the dominant use of these canals by the eighteenth century, mainly to support viticulture. By the mid-nineteenth century, viticulture was predominant, and the industrial use of the water disappeared completely (Ladki et al., 2012). With the abolition of feudalism, the canals became the property of independent groups of farmers, who maintained the pattern of collective maintenance and management of the canals. Through a variety of constitutional arrangements, the customs, laws, and rights governing the collective management of canals and irrigation remained fundamentally unchanged (de Passa, 1821: 261). The Civil Code adopted at the time of the French Revolution included an article calling for "maintenance of particular and local regulations on watercourses and the use of water" (de Passa, 1821: 262). The pattern of collectively owned and managed irrigation canals persisted through the constitutional adventures of the nineteenth century (Ruf, 2009). By the law of June 21,1865, the state codified these arrangements, creating the legal entity known as the Authorized Syndicated Association (ASA) and granting it the status of public agency or institution (établissement public) (Ladki et al., 2012: 29).

The Eastern Pyrénées remains among the French departments with the highest concentration of collective irrigation systems and with the highest concentration of ASAs. Currently there are some 3000 km of irrigation canals in the Department (Departement Pyrénées-Orientales, 2017). While much of the basin is equipped for irrigation, our study is focused on the valley of the main stem of the Têt between Vinça and Perpignan. This is the region that is most densely equipped for irrigation.

According to Thierry Ruf, water scarcity has always marked the political culture of the region:

> It is a culture marked by violence: water conflicts emerged shortly after the construction of canals. They had both a local and temporary nature and a regional and political character: conflicts of authority between powers have marked the Catalan history. To assert authority over water allocations is almost a custom . . . .
>
> (Ruf, 1998: 5)

This history of conflict extends to the present time; however, as we argue below, it has shifted from conflict between farmers and between ASAs – which we argue had positive social aspects – to a struggle between the state and the local irrigation authorities over water rights and over questions involving water use more generally.

Throughout most of the territory of modern France, there has been a strong correspondence between the control of rivers and the expansion and strengthening of the French state.[1] The territorialization of the modern state – its spatial effect by integrating and consolidating territory – was manifest in the promotion of rivers as axes of navigation and fluvial transportation (of wood) in the *ancien régime*, "demonstrating the application of a "modern" power, one of whose preoccupations was to create an economic space guaranteed by the State" (Serna, 1999: 33). This power was formalized and codified in law by, for example, by the Ordonnance des Eaux et Forêts of Colbert in 1669, which placed restrictions and controls on fishing, consistent with an overarching policy of promoting the use of rivers for transportation (Barthélémy, 2013: 38). From 1669, all watercourses capable of supporting navigation and transportation of wood were defined as state rivers (rivières domaniales), in the sense that these watercourses were brought under the authority of the king. To the rivières domaniales, in 1919, the state expanded its control over watercourses suitable for the production of hydro-electricity, "effectively extending the state's authority beyond navigable rivers to any that might produce electricity" (Pritchard, 2011: 44).

With the post-industrial economy and the ecological turn in the latter part of the twentieth century, the manner by which the French state intervenes in the nation's rivers changes: there is less emphasis on navigation, increasing decentralization of administration of the rivières domaniales, and decentralization of water management through the six metropolitan water agencies, which were established in 1964 on the scale of the large hydrographic basin. Today, in addition to traditional means of control, the state exerts control over rivers (and people) more indirectly through environmental regulation, and through the sticks and carrots of fining and subsidizing activities in support of reaching water quality objectives, standards of ecological continuity, and water-use efficiency.

The Eastern Pyrénées and local control of irrigation

The Eastern Pyrénées region either avoided or resisted these encroachments of the French state until very recently – especially with respect to the management of water. The political history of the region is relevant here: Prior to the Treaty of the Pyrénées in 1659 – by which the region was ceded to the French king by Spain – most of the present Department was part of the former Principality of Catalonia. Most inhabitants of the region have historically been Catalan-speaking, and it is still referred to by some as Northern Catalonia. The region was invaded by Spain in 1793 and only became definitively part of France the following year.

As noted above, Colbert's Code of Waters and Forests of of 1659 (the same year as the Treaty of the Pyrénées) put navigable rivers under the authority of the French king. According to T. Ruf, the Têt was originally classified as a rivière dominiale, which would have put the Têt under the authority of the French crown. However, owing to the region's particular political history, the legal regime of rights for rivers, and for water generally, remained somewhat "chaotic". At least until the French Revolution the ancient Catalonian "loi Stratae", was recognized, granting the right of use to landowners and long-long established communities (Ruf, 1998: 20).

Thus instead of coming under the direct control of the French state, as in most other parts of the country, water law and administration retained a strongly local character. This local character of water rights and water management is emphasized in the work of agronomist, historian, and political figure Jaubert de Passa (1785–1856). De Passa, who was from the region, was hired by the French government to study irrigation techniques in the Pyrénees Orientales and make recommendations. Consistent with the region's history and political culture, he advocated local rather than central management of water resources, a formula that was eventually adopted by the government in the form of the ASAs, described in more detail below (Mollard, 2001). The independent, local character of water management has persisted, manifesting itself in resistance to recent encroachment attempts of the state. For example, the regime by which state water agencies in France tax water use and provide subsidies to encourage efficiency – which has been in place since the 1960s – was rejected in the Eastern Pyrénées, citing the historical authorizations granted for agricultural water use dating back to the Middle Ages (Jacques Feraud 2017, pers.comm, May 2017). This resistance to state control of water management remains until the present; however it has been weakened as a result of the advent of pressurized irrigation technologies in the region, beginning in the 1980s.

Perhaps the most significant problem that these communities needed to resolve in order to practice agriculture in the Têt basin was – and is – what might be described as hydrological uncertainty. The hydrograph of the river presents typical seasonal fluctuations of a snow-dominated regime, with a low-flow period in July and August, when water is needed most for fruit-growing, market gardening, and forage, which dominate production in the region. At this time, the minimum three-consecutive-day volume can drop to 1.3 m^3/s, in the case of a five-year dry period. Farmers have thus had to develop means of securing access to water in times of low flows. The traditional technique they developed was a form of water-sharing that assured the fair distribution of scarce water resources, based on a water-tour rotation. Each farmer within a unit defined by a secondary canal was allocated water in proportion to the amount of land cultivated and the particular water requirements determined by different crops grown. The fairness of the system was – and where gravity systems are in place, still is – assured by an informal system of mutual surveillance, combined with informal, ad hoc arrangements between farmers to regulate differences in their annual water needs as a function of different crops grown and land under cultivation.

There has been a tendency, in political and popular culture in France, to regard gravity irrigation and the system of "tours d'eau" as antiquated, quaint, and even brutal. As Chantal Aspe has noted, the romanticized stories of feuding and fighting over water in the Mediterranean region are legion in French literature (Aspe, 2012: 10). However, the conflict that is often associated in the popular imagination with the water-tour system is often exaggerated, and in any case can be regarded as an impetus to the development and maintenance of a longstanding, successful tradition of collective management.

The social relations that evolved in order to deal with hydrological uncertainty are still present in the form of ASAs and traditions of local water management in the region. An ASA is a legal entity – a corporation – that brings together owners of adjacent properties to collectivize and coordinate the development, use, and maintenance of irrigation canals, and more recently of pressurized (including drip) irrigation systems. ASAs date from the mid-nineteenth century, since which time the French state encouraged local institutions of property-owners to collectivize the management of their common resources. Democratic structures (although the strength of members depends partly on water rights held), ASAs are led by members elected in general assemblies. Budgets are over-seen by the public treasury (Ruf, 1998: 15–16). In the Eastern Pyrénées, there has been a reduction in number, as smaller ASAs have consolidated into larger ones. In the early twentieth century, there were some 400 ASAs managing over 5000 km of canals for the irrigation of 30,000 hectares. Today there are about 220 ASAs, representing 3000 km of main canals and 30,000 hectares of irrigated land (including 6000 ha under pressure) in support of arboriculture, horticulture, and forage (Departement Pyrénées-Orientales, 2017). They range widely in size with the smallest including some 30 members and the largest including around 3000 members.

The relations between ASAs and the state are complex and often contradic-tory. The ASA is approved of by the state, which has attempted to use it to fur-ther particular objectives. Originally, with de Passa's recommendations, it served as a means of modernizing agriculture by virtue of collectivizing resources for rational canal development and maintenance. The process of collectivization however, produced its own dynamic, by which ASAs developed independently so as to resist the state's efforts to control water use in the region. Eventually, and especially with the advent of state financing of irrigation infrastructure begin-ning in the 1980s, some ASAs have become instrumentalized by the state to promote policies of water-use efficiency through pressurized irrigation systems.

The Vinça dam

A dam was built in the headwaters of the Têt basin (the Bouillouses dam) in the first decade of the twentieth century, capable of stocking 17 million m^3 of water. This dam was built and operated for generating electricity in the basin and had little effect on water supply for irrigation until an agreement was signed with the state electrical producer, Electricité de France, in the 1960s (Ruf, 1998: 9). Even then, water supplies were uncertain in the summertime, especially in the lower valley, which prompted the construction of a new dam. Imagined and first proposed in the early part of the century, the Vinça dam was built by the French state in the 1970s, at the tail end of the era of modern water. Its main purpose was to control floods and to ensure adequate water supply for irrigated agriculture. Today the dam struggles to contribute positively to the new, ecological standards of water management; it is judged to have been poorly designed, with its reservoir afflicted by high rates of sedimentation that the dam is incapable of discharging as well as poor water quality due to eutrophication and algae blooms.

The dam's main political proponent was Leon Jean Grégory (1909–1982), an influential figure, senator from 1948 to 1982 and mayor of Thuir from 1947 to 1982. Grégory had a vision to modernize agriculture in the region by providing reliable water resources by means of the dam at Vinça. He argued that the dam, by augmenting surface water irrigation via the canals, would improve summer groundwater recharge, especially further down the basin, where the larger population centres were located. His vision of "groundwater for drinking water, and surface water for agriculture" was consistent with the hydrogeology of the region and with the gravity-based irrigation techniques of the day. Grégory managed to put in place the necessary expropriations and state financing of the dam, which was finally completed in 1976, with a reservoir capacity of 24 million cubic meters.

Preliminary studies by the state (Génie Rurale et Ponts et Chausses) had posited a deficit between availability of water and the needs of farmers. In the inquiry prior to the D.U.P. (Declaration of Public Utility) dated August 27, 1970 the engineers stated:

> To illustrate the imbalance between needs and resources, we will quote two figures, one of 14 m^3 / second, corresponding to the water rights of the ASAs, and 5.5, 2.4, and $4m^3$ / second, corresponding to the average flows at Vinça in July, August and September, given the release of water made from the dam at Bouillouse, it being understood that the actual flow rates during drought significantly fall below these values.
>
> (Archives Départementales des Pyrénées-Orientales, Étude Barrage Vinça, Document Cote 1388w12, Borderau C)

In an era when supply management was the dominant paradigm for dealing with problems involving water availability (Gleick, 2000) this apparent (constructed) scarcity served as an argument for building the dam, in order to permit irrigation withdrawals at traditional rates throughout the summer months. Today, as discussed below, the same fact is interpreted as an argument for reducing farmers' use of water for irrigation.

At first, most farmers in the region either did not want or were indifferent to the construction of the dam. Archival records produced at a public inquiry for an earlier project to build the dam (inquiry of 1953–1955) reveal that contrary to what one might have expected, the farmers did not particularly want the dam. Based on these records and on interviews conducted in 2015 and 2016, we learned that farmers in the vicinity of Vinça were the strongest opponents of the dam because of the perceived impacts of the reservoir on their land, and because the benefit of assured water supplies would not accrue to them. But, as was evident in their testimony at the time, even the downstream farmers did not approve of the dam (Archives Départementales des Pyrénées-Orientales, Étude Barrage Vinça, cahiers de doléance 1953–1955). It is remarkable that so few farmers wanted the dam built, but perhaps not surprising, given the longstanding distrust of the state in this region, and given the concerns farmers continue to have about transferring authority for water management to the state. Underlying these concerns was the farmers' fear of having their traditional water rights

58 *Jamie Linton and Etienne Delay*

curtailed or limited. Further, having developed techniques to deal with hydrological uncertainty at the scale of the secondary canal, they were loath to change the entire network of social relations and the structure of authority brought about by shifting management to the basin-scale, as was implied by the dam.

Having managed hydrological uncertainty by themselves, through a combination of formal and informal rules, farmers were reluctant to move towards a system that produced certainty through dependence on state agencies, abstract calculations, and technologies that were out of their control. Having mastered uncertainty in their own way, farmers were wary about shifting to what was for them a new suite of uncertainties involving operation by technocrats of a large dam. While the dam would produce certainty of supply, thus ensuring a kind of water security, the farmers actually felt less secure with the dam because its management was out of their hands. According to the then-president of the Ile-Sur-Têt ASA in 1955, the downstream farmers preferred that the money spent on the dam be used to improve the canals and to build a network of small dams in the headwaters of the tributaries of the basin (ibid.). Eventually, however, the farmers were won over by Gregory, whom they held in high regard, and who reassured them that the dam would provide greater hydrological security while in no way threatening their water rights.

From gravity to pressure

The dam, completed in 1976, was financed by the state and owned and managed by the Department, which represents the state. The day-to-day management of the dam, however, was contracted with a private company, BRL (formerly la Compagnie Nationale d'Aménagement de la Région du Bas Rhône et du Languedoc). BRL proposed an alternative irrigation system now made possible by the certainty of water supply provided by the dam – a pressurized system at Thuir that would make spray and drip irrigation possible. This commodification of irrigation water, while consistent with the business plan of BRL, represented a radical departure from the shared resource hitherto managed by the ASAs. Pressurized irrigation was especially enticing to farmers because it meant a drastic reduction in labour in comparison with the traditional gravity techniques. However the fixed investments of this system were heavy and user fees were very high, resulting in failure of the enterprise. The local ASA (Thuir) denounced the contract with BRL and eventually overtook direct management of the system itself, reducing user fees by half (Ruf, 1998: 16–22). This has since become the model for pressurized systems throughout the basin – managed as separate systems by the ASAs.

Pressurized irrigation is associated with several important changes in the political economy and the social relations of agricultural production in the region. Pressurized (especially drip) irrigation is far more efficient in terms of water-use than gravity (furrow) irrigation. In addition to the traditional concerns about imbalance between supply and demand in summer, the new technology promised to resolve a variety of problems that have emerged since the Vinça dam was built, and that have become a growing priority for the state. With the passage of the Fishing Act of 1984, minimum streamflow requirements were first introduced into French law. Since then, in addition to a resource-scarcity problem, low flows in the Têt basin are regarded by the state as an environmental problem, heightened by the need to maintain European Water Framework Directive standards for good ecological status. Since the 1980s therefore, with the advent of environmental regulations to improve fish habitat, and with general shift from a supply-management to a demand-management paradigm of water management, the state has encouraged farmers to switch to drip irrigation by means of financial incentives offered to ASAs by the state water agencies. Thus beginning in the 1980s, ASAs began paying water taxes to the state for the first time, while receiving subsidies for canal works, which were now considered infrastructures necessary for the installation of pressurized irrigation (Jacques Feraud 2017, pers.comm, May 2017). We argue this has had the effect of changing the nature of the ASA. Where pressurized irrigation systems have been put in place, the ASA is transformed from a structure of collective resource management to an instrument of the state, through which taxes are collected and subsidies are paid to alter farmers' behavior.

Ruf (1998: 23) rightly points out that pressurized irrigation systems have the effect of reducing competition and internal tension within ASAs. However, as we have suggested, this internal tension may be regarded as salutary because it gives rise to formal and informal practices to assure fairness of distribution along secondary canals. As Aspe (2012) has pointed out in describing the social effects of irrigation dams in the region,

> . . . the development of dams has played an undeniable role in producing the amnesia of know-how and local knowledge of the hydrological system of rivers and altered the practices of managing scarcity . . . The technique of dams in regulating flows and water supply and in protecting against floods has erased centuries of history with surprising rapidity.
>
> (Aspe, 2012: 12)

Drip and spray irrigation has obvious benefits for farmers. It makes their life much easier. It does away with the need for nighttime irrigation, is less demanding physically, and requires less time to manage. The effect on individual farmers can thus be considered salutary. And yet, this shift from supply management (which motivated the building of the dam) to demand management has been less well received by farmers than might have been expected. While many ASAs – especially those located further up the basin, near the dam – have opted for pressurized systems on all or part of their territory, a much higher proportion of land in the lower basin remains under gravity (furrow) irrigation.

Why the reticence of farmers to adopt pressurized irrigation? We suggest there are two main reasons: The first relates to the different social relations that are structured by different types of irrigation. Gravity-based irrigation structures social relations in a linear sense, in that downstream farmers are dependent on

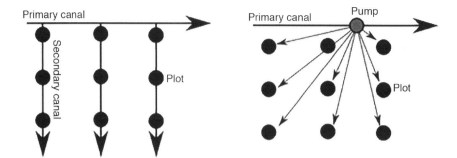

Figure 4.2 Two patterns of social relations structured by irrigation: The figure on the left represents the traditional gravity-based system, and on the right, a pressurized system. The dots represent individual plots.

upstream farmers to deliver water to them (Figure 4.2) People thus need to enter into agreement up and down the line in order to work out an arrangement for sharing water fairly. Gravity structures a linear set of dependencies, which is resolved by agreements allowing for farmers to take turns in watering up and down the line. This is a technology that seems to favour the development of convivial social relations. The architecture of agreement structured by gravity permits further informal agreement at smaller scales. For example, in our interviews with farmers we learned that it is not unusual for neighbours to make agreements among themselves: a farmer who needs more water than usual in a given tour might enter into agreement with another who needs less, to supply more water to the system the following tour in order to effect a quid pro quo (Mr. R. Majoral, president of the Thuir ASA, 2016, pers.comm, February 2016). Furthermore, the visible presence of water in the fields in a gravity system makes it possible for all parties to the agreement to assure themselves and each other that the agreement is being met simply by means of visual surveillance. Altogether, the principle of fairness is assured by informal and non-sophisticated methods.

Pressurized irrigation, by contrast, structures a different set of social relations, producing a different kind of water. Whereas the architecture of a gravity-based system is rather like a tree, a pressurized system is more like a star, with the pump at the center and individual consumption units radiating out. Pressure obviates the need for discussion up and down the line. Rather, everyone addresses him or herself as an individual to the center – to the pump, or rather to the manager or agency that operates the pump. The need for informal visual surveillance is obviated – indeed it is no longer possible – and a technician replaces this function with a water meter, with the technician now charged with the responsibility of distributing, and verifying, measured and metered quantities of water. These quantities of water, moreover, are sold per volume, effectively transforming the water from a shared resource into a commodity.

One reason why farmers in the Eastern Pyrénées have resisted pressurized irrigation is because they value the networks of social relations that are structured by gravity and are unwilling to relinquish these in exchange for a system has the effect of individuating farmers and transferring knowledge and control of irrigation technology from farmers to technicians.[2] As the president of the ASA of Thuir told us in an interview, "The dam has been the cause of a loss of public spirit (civisme) because before, we relied more on our neighbors for water" (Mr. R. Majoral, president of the Thuir ASA, 2016, pers.comm, February 2016). There is moreover a strong – and related – resistance to accessing water a commodity. As Ruf (1998: 23) observed after examining the irrigation situation in the Têt in the late 1990s, despite the actual and potential reduction of conflict between villages and neighbours made possible by pressurized systems, "nobody wants to pay for water in volumetric form and moreover to a third party, the State, the Regional Council or a company."

The resistance to commodifying water relates to another reason why farmers are reticent to adopt pressurized irrigation: because they are concerned this will result in reduction of their historic water rights. Farmers regard the state's promotion of demand management through pressurized irrigation as a strategy to deprive them of their ancestral rights to water. Here, it is relevant to note that for several years, state agencies have been promoting a plan to rationalize water use in the Têt basin known as the Plan pour la Gestion des Ressources en Eau (PGRE), or water resources management plan. A key component of this plan is the promotion of pressurized (drip) irrigation through incentives offered by the water agency. The main mechanism for pursuing this is through financial support of a mixed syndicate responsible for developing the PGRE. However, despite a variety of incentives, progress in developing the plan is currently stalled, mainly due to the farmers' distrust of the process, and their suspicions that the push to have them adopt greater water-use efficiency and pay for water will effectively deny them their water rights. The result, as noted above, is that despite all efforts by the state, a high proportion of the basin remains in a gravity-based regime: of 30,000 ha of irrigated land, only 6000 ha are under pressure.

Conclusion

Our study of irrigation in the lower Têt River basin is not at odds with what Aubriot (2013) describes as a "decryption of social relationships through the study of irrigation". This is consistent with that notion that the structure and practice of irrigation mirrors social relations (Aubriot, 2004). At the same time, with Aubriot (2013) we emphasize the dialectical position, which holds that while irrigation is in one sense a social construct, the material structure of an irrigation system equally affects social relations. Overall, we describe how the Vinça Dam and the advent of pressurized irrigation have contributed to an historical shift towards the production of what could be described as "modern water" (Linton, 2010) in the Eastern Pyrénées and the shift in social relations that this entails.

The Vinça Dam has produced the hydrological certainty whereby new actors, technologies, and patterns of social relations have come to characterize irrigation agriculture in the lower Têt basin in recent decades. The dam radically alters the type of expertise and the techniques for managing hydrological uncertainty, while transforming irrigation water from a shared, collectively managed resource to a commodity wherever pressurized irrigation has been established. While dependent on the dam in order to assure a standing reserve of water as a resource, pressurized irrigation structures a different set of hydro-social circumstances and produces a different kind of water than that entailed in gravity-based irrigation.

The farmers' resistance to pressure, we argue, is due to their reluctance to alter the social relations that are structured by gravity as well as their reluctance to pay for a commodified form of water that gets produced within the set of (individuating) social relations that form under pressure. Finally, the dam has had the effect of strengthening the territorial presence of the French state in a region that historically has avoided or resisted state control. This presence is realized directly by the transfer of water management authority from local ASAs to the experts who manage the dam, and indirectly, through the economic incentives applied by state agencies to influence the adoption of drip irrigation by farmers. The strengthening of the state in the region is balanced by weakening the collective associations and arrangements that farmers had long maintained in order to deal with hydrological uncertainty:

> At the end of the day, drip irrigation is a formidable instrument for disrupting social solidarities around water . . . Drip users no longer participate in collective efforts. The maintenance of surface networks useful to all do not motivate them anymore.
>
> (Ruf, 2009: 8)

This conclusion challenges the dominant approach to water security, which holds that this is best achieved through procuring certainty of supply. As Zeitoun et al. have pointed out, "the clearest research messages and policy recommendations currently on offer come from a 'security through certainty' stream . . ." (2016: 144), which "sees increased reliability of supply as a key component to water security in many contexts" (ibid.: 149). These and other researchers have called into question the presumption that farmers always benefit from more reliable and predictable supplies of water. To these critiques, we would add the notion that that hydrological uncertainty and insecurity of supply can actually yield social advantages and that hydrological certainty can bring some hidden social costs.

Notes

1 For a fuller account see Jamie Linton and Jacques-Aristide Perrin, "French Specificities of the Concept of River Continuity: History, Politics and Practices", forthcoming.
2 Ruf gives somewhat different reasons "why half the users have not switched to the pressurized watering system . . . difficulties of capitalisation in small operating structures, a fragmentation of farms that are too small for the equipment, a certain conservatism" (1998: 28).

References

Aspe, C. (2012) De l'eau agricole à l'eau environnementale. *In:* Aspe, C. (ed.) *De l'eau agricole à l'eau environnementale: Résistance et adaptation aux nouveaux enjeux de partage de l'eau en Méditerranée.* Paris: Éditions Quae.

Aubriot, O. (2004) *L'eau, miroir d'une société: Irrigation paysanne au Népal central.* Paris: CNRS EDITIONS.

Aubriot, O. (2013) De la matérialité de l'irrigation: Réflexions sur l'approche de recherche utilisée. *Journal des anthropologues,* 132–133, 123–144.

Banister, J. M. (2014) Are You Wittfogel or Against Him? Geophilosophy, Hydro-Sociality, and the State. *Geoforum,* 57, 205–214.

Barthélémy, C. (2013) *La pêche amateur au fil du Rhône et de l'histoire: usages, savoirs et gestions de la nature.* Paris: L'Harmattan.

Bichsel, C. (2016) Water and the (Infra-)Structure of Political Rule: A Synthesis. *Water Alternatives,* 9(2), 356–372.

De Passa, J. (1821) *Mémoire sur les cours d'eau et les caneaux d'arrosage des Pyrénées-Orientales, par M. Jaubert de Passa.* Paris: Huzard.

Departement Pyrénées-Orientales. (2017) *Les canaux d'irrigtion* [Online]. Available: www.ledepartement66.fr/391-les-canaux-d-irrigation.htm [Accessed May 13, 2017].

Gleick, P. H. (2000) The Changing Water Paradigm: A Look at Twenty-First Century Water Resources Development. *Water International,* 25(1), 127–138.

Ladki, M., Guérin-Schneider, L., Garin, P. and Baudequin, D. (2012) Des canaux d'irrigation aux canaux de distribution d'eau brute ? Regarder le passé pour comprendre le présent et préparer l'avenir. *In:* Aspe, C. (ed.) *De l'eau agricole à l'eau environnementale: Résistance et adaptation aux nouveaux enjeux de partage de l'eau en Méditerranée.* Paris: Éditions Quae.

Linton, J. (2010) *What Is Water? The History of a Modern Abstraction.* Vancouver, British Columbia: UBC Press.

Linton, J. (2014) Modern Water and Its Discontents: A History of Hydrosocial Renewal. *WIRES Water,* 1(1), 100–111.

Linton, J. and Budds, J. (2014) The Hydrosocial Cycle: Defining and Mobilizing a Relational-Dialectical Approach to Water. *Geoforum,* 57, 170–180.

Mollard, E. (2001) "Recherches sur les arrosages chez les peuples anciens" de F.J. Jaubert de Passa (1846): une histoire de la gouvernance avant l'avènement de la technocratie. *In:* Thierry, R. and Rivière-Honegger, A. (eds.) La gestion sociale de l'eau, concepts, méthodes et applications. *Territoires en Mutations,* 2004(12), 17–31. ISSN 1278–4249.

Obertreis, J., Moss, T., Mollinga, P. P. and Bichsel, C. (2016) Water, Infrastructure and Political Rule: Introduction to the Special Issue. *Water Alternatives,* 9(2), 168–181.

Ostrom, E. (1992) *Crafting Institutinos for Self-Governing Irrigation Systems.* San Francisco: ICS Press.

Pritchard, S. B. (2011) *Confluence: The Nature of Technology and the Remaking of the Rhone.* Cambridge, MA: Harvard University Press.

Ruf, T. (1998) *Gestion collective de l'eau dns la vallée de la Têt (Pyrénées Orientales).* ORSTOM, Agricultures irriguées durables. Groupe de travail IRRI-Mieux, Gestoin collective d'une ressource commune l'eau.

Ruf, T. (2009) *A propos de l'Avenir de l'eau d'Erik Orsenna : La caricature des eaux* [Online]. Available: www.eauxglacees.com/IMG/pdf/Critique_Orsenna-V5.pdf [Accessed May 1, 2017].

Ruf, T. (2011) Le façonnage des institutions d'irrigation au XXe siècle, selon les principes d'Elinor Ostrom, est-il encore pertinent en 2010? *Natures Sciences Sociétés.* 19(4), 395–404.

64   *Jamie Linton and Etienne Delay*

Serna, Virginie (1999) Le fleuve de papier. Visites de rivières et cartographies de fleuve (XIIIe-XVIIIe siècles). *Médiévales,* 36, 31–41.

Swyngedouw, E. (1999) Modernity and Hybridity: Nature, *Regeneracionismo*, and the Production of the Spanish Waterscape, 1890–1930. *Annals of the Association of American Geographers*, 89(3), 443–465.

Swyngedouw, E. (2004) *Social Power and the Urbanization of Water: Flows of Power.* Oxford: Oxford University Press.

Wittfogel, K. A. (1957) *Oriental Despotism: A Comparative Study of Total Power.* New Haven: Yale University Press.

Worster, D. (1985) *Rivers of Empire: Water, Aridity, and the Growth of the American West.* New York: Pantheon Books.

Zeitoun, M., Lankford, B., Kreuger, T., Forsyth, T., Carter, R., Hoekstra, A. Y., Taylor, R., Varis, O., Cleaver, F., Boelens, R., Swatuk, L., Tickner, D., Scott, C. A., Mirumachi, N. and Matthews, N. (2016) Reductionist and Integrative Approaches to Complex Water Security Challenges. *Global Environmental Change*, 39(143–154).

5 Big projects, strong states?

Large-scale investments in irrigation and state formation in the Beles Valley, Ethiopia

Emanuele Fantini, Tesfaye Muluneh and Hermen Smit

"Welcome to the Meles Zenawi Headquarters". The signpost at the entrance of the Beles Sugar Development Project's offices powerfully recalls how water management dovetails materially and symbolically with the Ethiopian government's strategy of state building and economic growth. Over the past decade Ethiopia has embarked on a series of big projects for hydropower and irrigation development, with the goal of achieving the status of "carbon free" middle-income country by 2025 (FDRE, 2011). The central federal government and public enterprises retain a primary role in the design and implementation of these projects, in line with the tenets of the developmental state, as carved by its main ideologue and architect, the late Ethiopian Prime Minister Meles Zenawi (Zenawi, 2012).

This chapter analyses the relations between water management, large-scale land acquisitions and state building in contemporary Ethiopia. In doing so, we adopt a dynamic understanding of statehood, based on Bruce Berman and John Lonsdale's (1992) distinction between *state building* and *state formation*. The latter is "the historical process, mainly unconscious and contradictory, of conflicts, negotiation and compromises between different groups" (Berman and Lonsdale, 1992: 5) that occurs within – and ultimately shapes – the explicit institutional strategies of state building promoted by governments' elites. By focusing on the consequences – more or less unintended – of public authorities' institutional strategies, as well as on their negotiation or contestation by different groups (Hagmann and Péclard, 2010), we acknowledge that "the state is always in the making" (Lund, 2016: 1199). This fluid understanding of the state suggests an analysis of big development projects both as a means and as an effect of state building: large infrastructures can cement the state's capacity to control people and resources; at the same time they are also the expression and manifestation of such power, as well as of its limits.

Big development projects are not new in Ethiopia. Throughout different political regimes – Imperial Ethiopia (up to 1974), the filo-Soviet military regime of the *Derg* (1974–1991) and the contemporary ethno-federal republic (since 1991) – big projects played a key role in state-building strategies and in the processes of resources control, people *encadrement* – namely "incorporation in structures of control" (Clapham, 2002: 14) – and exercise of power. We use the French word *encadrement* since it refers simultaneously to the frame – in our case irrigation as key element for agriculture modernisation and economic growth – and to the process of framing – here the reconfiguration of landscape and the reordering of society by dint of the irrigation scheme.

Over the past five years the revamping of big projects to sustain unprecedented figures of economic growth has led to a renewed attention by scholars, policy makers, international organisations and NGOs. Large-scale investments in agriculture have been analysed mainly within the broader debate on land grabbing, highlighting how in the Ethiopian case large-scale land acquisitions do not result in the erosion of national sovereignty as it is often assumed. On the contrary, in Ethiopia the central government plays a leading role in these investments, with the goal of building its presence in the country's peripheries, enforcing political authority and exercising control over peoples and resources (Moreda, 2017; Lavers, 2016). These projects represent a new chapter of the longstanding centre-periphery relations in the process of state formation in Ethiopia. Here the periphery is defined not merely in geographic terms, but by its "marginal position in the power structure of the state", as well as by "the loss of native resources appropriated by the state and transferred to the centre, economic exploitation and cultural discrimination" (Markakis, 2011: 7). James Scott's framework on high modernism (Scott, 1998) has also been evoked to highlight the link between large-scale development projects – particularly big dams – and the authoritarian exercise of political power in contemporary Ethiopia (Mosley and Watson, 2016; Fantini and Puddu, 2016).

We wish to contribute to this debate by taking as case study the Beles Sugar Development Project (hereafter the Beles Project), a large-scale investment in irrigation implemented by the state-owned Ethiopian Sugar Corporation (ESC) which stretches across the border of the Amhara and Benishangul-Gumuz regions. Our analysis draws on field research conducted between September 2015 and March 2016[1] and focuses on two main elements. First is the process of state territorialisation, or the discursive and material production of the state's territory through the restructuring of the biophysical space. This process entails the making of a new centre of investment, through considerable flows of financial, natural and human resources, in an area traditionally considered a periphery of the country. Second, the transformation and continuities in the socio-spatial reordering of people and resources at the margins of the new centre, by techniques such as resettlement, and their consequences in terms of subject formation, mobility, transactions and (re)distribution of land, water and labour. From this perspective, the Beles Valley still remains a periphery of extraction.

The combined effect of these dynamics suggests a major contradiction in the process of state formation in contemporary Ethiopia. On the one side, state territorialisation through the monoculture of sugar cane allows the central federal government to increase its capacity in terms of resource extraction and accumulation at the periphery of the country. This process of concentration and homogenization aims at rendering the biophysical space and the population more "legible" (Scott, 1998) and exploitable to sustain economic growth. On the other side, the same process engenders at its margins, or in its interstices, a plurality of patterns of transaction and resource redistribution, resulting in increasing social and spatial mobility. The reconfiguration of the relations between people, spaces, stakes and resources increases the complexity of the society and makes it less intelligible to the Ethiopian government's rather rigid ideology and strategy of people *encadrement*. Thus, while reinforcing the central government's capacity to control resources, the making of the Beles Project might at the same time jeopardise the legitimacy of its political authority.

The developmental state and sugar production

This section introduces our case study, recalling how sugar production is framed as key component of the government's official discourse on state building and economic growth, but also describing how this "sweet vision" (Kamski, 2016) contrasts with the rather sour reality of the ongoing implementation of the Beles and other sugar projects.

Sugar production is a key component of the development strategy of Ethiopia's Growth and Transformation Plans (GTP I 2010/2011–2014/15 and II 2015/16–2019/20) to strengthen the national economy and further integrate it in the global markets. Ethiopia aims to produce 4.9 million tons of sugar by 2020 to become a major sugar and ethanol exporter, to create jobs in order to address rapid demographic growth and youth unemployment, and to meet the growing internal demand for sugar. At the moment the country consumes up to 700,000 tons of sugar each year while the annual production is 440,000 tons. The sale of sugar in the domestic market is regulated and subsidised by the state and managed by the ESC. People are entitled to buy limited amounts of sugar at a subsidised price both for domestic consumption and commercial activities; they purchase it directly at small kiosks run by consumer unions or on local authorities' premises. Created in 2010 to replace the Ethiopian Sugar Development Agency, the ESC holds the monopoly on sugar production and processing in the country. To confirm the strategic role accorded to sugar production within the government's development strategy, the ESC director general initially enjoyed the rank of minister and reported directly to the Prime Minister's office. This position has been held until 2013 by Abay Tsehaye, one of the founders of the Tigray People Liberation Front (TPLF) – the most influential party within the ruling coalition of the Ethiopian People Revolutionary Democratic Front (EPRDF) – and former Minister of Federal Affairs and security advisor to the late Prime Minister Meles Zenawi. After a restructuring in 2015 the ESC director lost the rank of minister and is now accountable to the Minister of Public Enterprises (Kamski, 2016).

The expansion of the sugar industry is one of the Ethiopian government's flagship development programs that in terms of material and symbolic investments can be compared to the construction of big dams for hydropower. To reach its ambitious targets, the ESC has embarked on the expansion of existing factories (Wonji/Shewa, Metehara and Fincha sugar factories in the Awash and

Blue Nile river basins), as well as in the construction of ten new factories in the traditional peripheries of the country, such as the Tendaho Sugar Project in the Awash basin (Afar region), the Kuraz Sugar Project in the Omo-Turkana basin (Southern Nations Nationalities and People Region), and the Beles Project in the Blue Nile basin (Amhara and Benishangul-Gumuz regions). In the government's discourse, these areas are considered endowed with rich natural resources – particularly land and water – perceived as untapped or under-utilised because of the lifestyles and systems of production adopted by the local population (Rahmato, 2011) – mostly pastoralists like in South Omo or those practising shifting cultivation like in Benishangul-Gumuz – often labelled as "backwards".[2] In these frontiers the developmental state is vested with the mission of promoting agricultural expansion, fertility and modernisation through the introduction of technology, standardised systems of production and bureaucratic institutions. These ideas are conveyed by the notion of *lemat* – the translation of "development" in Amharic (Berhane-Selassie, 2009), the political language of state building throughout different Ethiopian regimes.

The Beles Project is located in the country western lowlands, at the border between the Amhara and Benishangul-Gumuz regions, the latter being the region where the Grand Ethiopian Renaissance Dam is also being built (Figure 5.1).

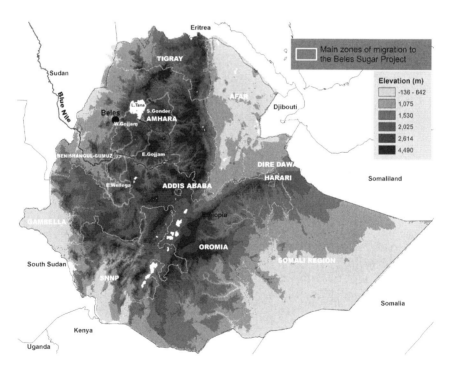

Figure 5.1 Location of the Beles Sugar Development Project and the main zones of labour migration towards it

(Credit: Tesfaye Muluneh)

The project is part of the Tana–Beles Sub-Basin, one of the five national growth corridors designed to enhance agricultural productivity, commercialisation and diversification. Launched in 2013, the project initially foresaw a scheme of 75,000 hectares irrigated with the water released from the Beles River, a tributary of the Blue Nile. The construction of three sugar factories was planned in order to produce 242,000 tons of sugar and 20,827 cubic meters of ethanol per year. Once fully operational, this investment was expected to create around 40,000 new jobs.[3] A feasibility study identified as the most suitable option a sugar estate of 36,000 hectares with the rest of the surface allocated to smallholder farmers and commercial farms (Clark et al., 2013). This solution was identified as the best compromise in terms of employment creation, minimising resettlement and avoiding the rise of food security issues. However the ESC decided to allocate the entire area to the sugar estate, in spite of identified higher financial and resettlement costs, lower equity and occupancy rates and risks of food insecurity (Clark et al., 2013). Therefore no private sugar cane outgrowers are present in the scheme, unlike in other ESC plantations such as Wonji or Tendaho. Other organisations involved in the project are MetEC (Metal Engineering Corporation), controlled by the Ministry of Defence, to whom the design and construction of the three factories was subcontracted, and the Amhara Water Works and Construction Enterprise and the Amhara Design and Supervision enterprise, controlled by the Amhara regional state and in charge of design and construction of the irrigation infrastructure.

In March 2016, three years after the launching of the project, approximately 20,000 hectares of land had already been cleared, and most of the sugar cane planted on the initial 13,000 hectares irrigated was ready to be processed. However, the first factory was still under construction and the works were only about to begin at the second factory's site. According to the Beles Project manager, the delay in the construction of the factories costs the Ethiopian treasury 7 billion birr (approximately 250 million euro) per year.[4] If the sugar cane is not processed on time, the sugar content in the cane begins to decrease and yields fall drastically. In the worst scenario, ESC will have to burn the sugar cane because it is not economically viable to transport and process it. This delay casts doubt on the future of the project. With similar delays at other projects, it also contributes to wider concerns about the sustainability and effectiveness of the overall strategy of agricultural and industrial modernisation through sugar cane production (Kamski, 2016). In May 2016 the CEO of ESC admitted to the parliament that two years after the completion of GTP I only one factory out of ten was almost completed, while 90% of the full payment had already been made to MetEC.[5] In November 2016 a US$1 billion agreement was announced with the Turkish group Bedisa to acquire a 75% stake in the Beles Project, probably in an attempt to speed up its implementation.[6] In January 2017 the Amhara Professionals Union, a group in the Ethiopian diaspora based in Washington launched an online petition, qualifying this investment as "land grab" and asking Bedisa management to withdraw it.[7] After an initial visit to the project site by Bedisa in the fall of 2016,[8] their involvement in the project does not seem to have materialised. According to the latest information displayed on the ESC website, the project has been downscaled to 50,000 hectares and two processing factories instead of three.[9]

A new centre: workers' narratives and expectations

The Beles Project being still under development and construction, it is premature to assess its overall impact and speculate on its social, economic and environmental sustainability. We found it meaningful, however, to examine the making of this project to better understand the enactment of the state's presence and power in the area. Indeed the Beles Project is the most striking manifestation of the state in this area: when describing its activities, achievements and consequences people usually refer to the project as being *mengist*, the Amharic word designating at the same time "the government, the party, the state, and all their agents" (Lefort, 2007: 256).

The project entails a vast process of state territorialisation, (re)producing political authority by transforming the landscape, standardising the space, accumulating and concentrating resources. The creation of the sugar scheme implies land expropriation and resettlement of smallholder farmers. According to the official statistics of the ESC, a total of 20,000 people will be affected. The landscape is being bulldozed and rendered homogeneous by seemingly never-ending sugar cane plants. New infrastructures are built: the canal and irrigation system based on gravity bringing water from the Beles river to the sugar plantation; the first sugar factory and the offices in the ESC headquarters, the workers' camps, the electricity grid and a new asphalt road crossing the whole sugar scheme and connecting the town of Fendika with Pawi *woreda* in Benishangul-Gumuz region. Thus a new centre is in the making within and around the Beles Project. The town of Fendika is growing rapidly in order to meet the demand for services and goods related to the implementation of the project.

The process of territorialisation implies relevant flows of capital, technology and labour to the area. Here we focus on the latter, since one of the main objectives of the project is to create jobs to address the employment demands of a growing adolescent population. Officially unemployment is described mostly as urban phenomena affecting in particular the educated youth. However, popular perceptions about unemployment are much wider. In rural areas a young man is vernacularly considered "unemployed" if he cannot become independent and create his own household because he does not work, or he does not earn enough, or he does not have enough land. A vast proportion of young men aged sixteen to thirty find themselves in this situation, with state authorities both at national and local levels fearing that they cannot control this "rural lumpen proletariat" (Lefort, 2015). The Beles Project is part of the strategy of the Ethiopian government to legitimise its authority in front of the youth, by addressing their needs and aspirations through economic diversification, industrialisation and jobs creation. Most of the 40,000 jobs that the project is set to create are reserved to skilled labourers and attract mainly youths coming from outside the area.

Big projects, strong states?   71

By November 2015, more than 1400 skilled staff were employed by the project (engineers, agronomists, technicians, foremen, support staff). These highly educated workers usually hold a university degree in engineering or agronomy. They are recruited from all over the country. For many of them, working in the Beles Project represents the first step in their professional career, an opportunity to get further training and to acquire practical knowledge and experience, as testified by a young irrigation engineer:

> After graduation I used to work in METEC for four months as hydraulic engineer. Then I was recruited by the ESC. We had training for six months. The first month at Bishoftu was mainly about politics, social life, and the Kaizen method. Then we had four months of theoretical training at Wonji sugar factory on irrigation, followed by written exam. Finally we had on-the-job training on different kinds of irrigation systems. We had also training at Tendaho sugar project for one month to get exposed to the remaining types of irrigation.[10]

Another ESC plantation expert, holding a diploma in natural resource management, emphasises the financial benefits of being employed on the project:

> While I was serving in Wenbera *woreda* as development extension agent my gross salary was only 1200 birr per month. I was involved in all kind of activities related to agriculture, without any opportunity for specialisation. Here my salary has grown incredibly: there is the benefit package, the food allowance of 120 birr per day, the remoteness allowance of 30% of the salary. So I get a net monthly payment of 6400 birr. For me this is a lot of money![11]

Because of the benefits and opportunities that these skilled workers get from the project, many of them are eager to go by and repeat the government's official discourse in terms of the project's contribution both to the national economy and to local development. One of them describes the project as follows:

> The project has huge benefits for the development of the country's economy. It provides employment opportunities and career development.[12]
>
> One of the main benefits of the sugar project is that it is source of foreign exchange for the country.[13]
>
> The project had a major impact on the growth of the *woreda* and of the town. In the past there were many problems. Now the town got drinking water supply systems, an asphalt road is under construction. Also schools and health centres have been constructed for the people displaced by the project. They used to live in the forest but now schools and drinking water have been made available for their children.[14]

Similar narratives emphasising the positive impact of the project in terms of local development are adopted by people migrating to this area and working in

72   *Emanuele Fantini, Tesfaye Muluneh and Hermen Smit*

the services sector that accompanies the implementation of the project and the expansion of Fendika town.

Beside qualified engineers and agronomists, in November 2015 more than 11,000 wage labourers were employed in the sugar scheme.[15] These workers are mainly recruited in the Amhara region, particularly in the highlands zone of South Gonder and West Gojam where limited access to fertile land makes it very difficult for youth to become independent and to form their own families by working as smallholder farmers. Local authorities, in collaboration with ESC officials, organise information campaigns and orientation sessions to recruit labourers; they select them and arrange for their transport to the sugar scheme. Educated and landless youth are the main target. To be recruited they need to have completed at least grade 10 of the national education system. Some of the recruited workers are also graduates from TVET (technical and vocational education training) or hold a diploma (10+3 on the Ethiopian education scale). Most of the workers have been attracted by the promise of a permanent job after an initial trial period, as recalled by one of them:

> I completed tenth grade in 2011 and thereafter lived jobless for three years with my parents. In February 2014, we – I think we were around seventy – were registered by the *kebele*[16] and they told us that we were going to get a job in the Sugar Project. They promised us that we were going to be shifted to permanent positions after three months.[17]

However in most of the cases the situation that workers found in the project did not meet their expectations. The most common complaints heard in the interviews are related to the hard working conditions in the scheme and their health consequences, the low salary and the poor facilities offered by the project.

> My feet swell most of the time, and my skin is always itching. I have already cut many times my fingers. Look at my nails! In the scheme we also see many different wild animals, like snakes, every day.[18]
>
> Everyday we work from 7 am to 3 pm. But there is no drinking water in the sugar scheme, and the weather is very hot. So most labourers drink the water used to irrigate the cane. Gastritis, malaria and typhus fever are common illness among the workers.[19]

Labourers are paid on a piecework basis. Therefore their monthly income fluctuates with the seasonality of the work and the different tasks that workers are asked to perform, such as cleaning the land, planting seeds or moving the irrigation sprinklers.

> Sometimes we do not have work and we stand idle the whole day. At the end there are months during which we cannot even cover our meal expenses.[20]

Workers also complain about the services and the facilities offered by the project: poor housing quality in the camps, lack of regular transport to the working sites, and lack of proper equipment and tools to protect themselves from the hardship of the job in the sugar scheme.

> At the beginning I used to live in the camp. The food provided there was not good and the rent was expensive. I could not save any money at the end of the month. Later, I moved to town to a rental house to live with my wife. I met her here. She is also a tenth grade graduate working as wage labourer. She came from Este, South Gondar. We arrived here on the same date, but from different *woreda*.[21]

Marrying a co-worker seems a recurrent strategy to cope with the costs and the hardship of making a new life in Fendika. It also looks like the first step towards meeting the aspiration to autonomy, independence and adulthood that pushed most of these young workers to join the Beles Project.

Labourers try to address their concerns and complaints through the Kaizen system that structures their daily work. Kaizen – the Japanese word for "continuous improvement" – is a business and management model that the Ethiopian government has borrowed from Japan to restructure the labour organisation within public institutions and companies (Fourie, 2015). Workers are organised in teams of five with one representative. They meet every week to monitor the work, discuss problems and identify potential solutions, later reported by the team representative to the project management. As explained by a worker,

> We write down ideas, opinions, needs and we sign the paper and submit it. For instance, we made requests about health care, medical services, and insurance in case of accidents. We asked gloves and we got them. We also raised the issue of payment delays. They told us that the order of payment for each team is based on a lottery system and that it was a matter of chance. But later they improved it and we started to get our payment on time.[22]

Interestingly, the Kaizen system seems to mirror the organisation of the so-called "development armies", the structures for people *encadrement* introduced by the EPRDF in the last decade. Different and overlapping groups of the population, such as farmers, youth and women, are organised in groups of five led by a peer leader, who is a member of the ruling party structure and acting as "role model" in monitoring the implementation of official national development policies down to the household level (Villanucci and Fantini, 2016; Lefort, 2012). Workers tend to interpret the Kaizen system as a device similar to the development armies, in spite of the very different theories and approaches behind them. For many who find their personal and economic aspirations frustrated, this system has become the face of a state whose official discourse of economic growth, modernisation and job creation continues to raise high expectations but has brought them very little.

Reordering an old periphery

Big development projects and related flows of people and resources are not entirely new in the Beles valley. This area has been historically perceived as a frontier endowed with natural resources – particularly water and land – up for grabs to decongest the Amhara highlands, the old core of the Ethiopian state, and to project its political influence over the peripheries of the country. Migration from the highlands of Gonder and Gojjam (Amhara region) or from Wallega (Oromia region) towards the lowlands of Metekel has been going on for centuries, pushing the autochthonous Gumuz population ever further westwards (Markakis, 2011). In the nineteenth century the Agaw, a group describing themselves as coming from the central north of the country, were forced to descend the Ethiopian plateau and ruled the Gumuz on behalf of the Amhara (Ficquet and Feyissa, 2015). Under the Ethiopian federal regime, the Agaw have been bestowed a self-government zone in Awi *woreda* within the Amhara regional state. However the continuous flow of highlanders from Amhara and Oromia to the lowlands of Metekel over the past few decades has pushed the Agaw into the margins of the zone bordering the sugar scheme (Wolde Selassie, 2009a).

Besides the large influx of migrants whose movement was not coordinated by the government, this area has been (re)shaped by the juxtaposition of different big development interventions and resettlement schemes. In 1985 the *Derg* government started the Tana Beles Resettlement project, entailing the transfer of 80,000 people from "overpopulated" and drought-affected highlands in Amhara (Sehwa and Wollo) and Tigray, towards 250,000 hectares of what was considered "unoccupied land with agriculture potential" in the Beles valley (Wolde Selassie, 2009b; Dieci and Viezzoli, 1992). In 2003 the EPRDF implemented a new round of voluntary resettlement from the Amhara highlands in the name of humanitarian concerns, food security and the fight against poverty (Pankhurst and Piguet, 2009). More recently, in 2008 people were resettled in the Beles valley from several villages in eastern Wallega (Oromia region) to address an ethnic conflict between Gumuz and Oromo (Markakis, 2011). The juxtaposition of different layers of migration and government-led resettlement institutionalized a system of smallholder farmers relying on rain-fed subsistence agriculture typical of the highlands, resulting in the marginalisation of the Gumuz and the Agaw, who used these lands for shifting cultivation.

The Beles Project adds a new layer of complexity and spatial reordering in the area, through the resettlement of almost 20,000 people in Awi and Metekel zones,[23] this time in the name of economic growth and agricultural commercialisation. The new claims on the area for irrigated agriculture adds to the shortage of land, with a significant number of farmers dispossessed from their means of livelihood in an area previously considered as endowed with abundant natural resources. Land in Ethiopia belongs to the state and it is leased to farmers that can transfer their usage right to their heirs. The process of land eviction, compensation and resettlement to create space for the sugar scheme has been undertaken by the local *woreda* (district) authorities with the support of

ESC. Local authorities are trapped in the difficult position of having to mediate between the instructions of the federal government and the claims made by the local communities.

Indeed, the resettlement generated many complaints about its fairness and unpredictability. The most critical issues in terms of impact on people's livelihood are the size, the quality and the location of the land received as compensation. The official size of land leased to a household in the area is 1.5 hectares, but most of the farmers whose land was taken by the Beles Project used to have plots of at least 5 hectares, secured by informally clearing and occupying unused land or by renting additional land. However, people who were registered as residents of the area were compensated only for the 1.5 hectares in the records of local authorities. Those who migrated from the highlands recently and are not registered as residents did not get any compensation. A second grievance is related to the quality of the land received as compensation: the rocky soil found in the new plots does not obtain the same yields of the old plots or cultivate the same varieties of crops. Third, farmers reported that the new land they were assigned is too far away – often one day's walking distance from their home – and sometimes already used informally by other people for farming or communal pasture. This fuelled several conflicts between the resettled farmers and those who were already informally using the land given as compensation. Common grazing land was not included in the compensation scheme either. Thus among the consequences of the new situation of land shortage there is also the reduction of the pasture and therefore of the cattle population.

People also complain about the unfair valuation of the assets, such as houses, wells or fruit trees that had to be abandoned and destroyed. The unfair valuation of assets resulted in the deterioration of housing standards and living conditions in the resettlement area, as described by a resettled farmer:

> I received 30,000 ETB as compensation for my house, which was not sufficient to build a new one. I had to sell four heads of cattle to cover the costs of the new house. And it is not completed yet. Furthermore, during the harvest season I have to move with my family in a *kenda* [temporary shelter] near my land. This is not an acceptable life. We accept it only because this is what the *mengist* has brought to us.[24]

Moreover, the resettlement process has been characterised by great uncertainty. For instance, people in Wombelase *woreda*, where the sugar scheme has not expanded yet, were informed four years ago that they had to move. As of November 2015, however, they were still waiting to know when and where they would be resettled. In the meantime, farmers had already stopped investing in their land, for instance by planting fruit trees or adopting soil conservation measures. One of the main consequences of these shortcomings is the increase in food insecurity: according to the data obtained from the local *woreda*, in 2015, 350 of the resettled households (approximately 1300 people) had to rely on food aid.[25]

The strategies of the resettled farmers to cope with the new situation led to increased social stratification and spatial mobility around the sugar scheme. Most of the farmers rent out the land they got as compensation, as they consider it to be too far away and not productive enough. As an alternative to farming, they try to join the Beles Project as wage labourers. However they often complain that the jobs in the sugar scheme are reserved for the "educated" ones, with most of the resettled farmers ending up in temporary and low-paid positions such as guards. The most resourceful of them try to take advantage of the expansion of the town and its economy by investing in new activities and occupations like transport of goods or waste collection and recycling. The majority of the resettled people still see agriculture as the preferred option. Others turn their eyes further westwards, towards Benishangul-Gumuz – or *Killil* (regional state) 6 as it is often referred to – where land is considered to be still abundantly available. Once again, people are on the move to rent land or sharecrop it. Many of them migrate temporarily during the harvest season, sometimes bringing their families, and in other cases relying on relatives to take care of them.

The making of the Beles Project thus reorders the old periphery of the state by further complicating tensions and inequalities in access to resources, between and within different groups. Newcomers from the highlands are attracted by job opportunities within the project and in the related services. Smallholder farmers that arrived in this area during past waves of migration were resettled at the margins of the project and try to cope with the new situation by diversifying their economic activities and transactions. Autochthonous groups like the Gumuz and Agaw are further marginalised: they were not compensated for the common grazing land or the farming land informally cleared. Mostly unable to get employment with the Beles Project, they search for new opportunities by migrating to neighbouring areas and relying on family networks for sharecropping. These processes also transform genders and generational relations. For instance, women take up additional roles and engage in new economic activities while their husbands are away for the harvest season. Farmers who lost their land – and sometimes their social role and identity along with it – become dependent on better-off relatives through land renting or sharecropping.

High modernist schemes have been traditionally interpreted – from James Scott onwards – as a way to make the territory and the population more legible to the state. However the social stratification and the spatial mobility engendered by the making of the Beles Project seems to increase the complexity of the society and its networks of relations, rendering a significant portion of the population even more invisible and less intelligible to the rather rigid EPRDF ideology and practices of *encadrement*, such as the development armies.

Furthermore, the fact that resettled farmers were compensated only in the measure that they were legible trough the local cadastre confirms that, as Lund has it:

> there is only one thing worse than being seen by political authority, and that is not being seen. . . . The processes of recognition of claims to land

and other resources as property, and of political identity as citizenship with entitlements, simultaneously invest the institution that provides such recognition with recognition of its authority to do so.

(Lund, 2016: 1205)

The stories presented here tell us how the rights and claims of the groups involved in or affected by the Beles Project enjoy different degrees of government recognition, and consequently how these groups invest the state with different degrees of legitimate political authority.

Conclusion

The making of the Beles Project shows us how the relevance of large-scale state-led irrigation stretches beyond agricultural production as it becomes a repository of symbolic and material meanings about the state for the institutions and the people implicated in the project. Sugar production being a cornerstone of the Ethiopian official strategy of state-led economic growth, the government has poured tremendous efforts into the Beles Project in terms of natural, human, financial and technical resources. These resources have been concentrated in the Beles valley to transform the territory of an old periphery into a new centre. The process has been accompanied and legitimised by an official discourse on economic growth, modernisation and jobs creation that contributes to creating expectations and aspirations within the different groups of people involved in the project. The shortcomings in the implementation witnessed so far cast severe doubts on the capacity of the project to meet all these expectations, and more broadly to uphold the political legitimacy of the government in the area. While many of the engineers and agronomists benefiting from the project in terms of income and career development reproduce the government's official discourse of state-led development, in the eyes of the majority of the daily workers in the scheme and of the resettled people, the project assumes the face of the authoritarian and unpredictable *mengist*.

Over the last ten years Ethiopia has witnessed an unprecedented pattern of economic growth, but also largely unexpected waves of popular protest and civil unrest, animated by farmers and students in different parts of Oromia and Amhara regional states. These protests have not yet burst into the open in the Beles valley – at least to our knowledge. However we believe that the making of the Beles Project sheds some light on how these phenomena – economic growth and popular discontent – might be related. The shift from a social fabric based on smallholder subsistence farming towards a more "modern" and stratified economic system of agriculture commercialisation is mortgaged on the highly uneven accumulation of resources and social exclusion. These processes tend to reproduce inequalities between different groups – national and local elites, highlanders and lowlanders, resettled farmers and educated workers – in terms of being recognised by the state and getting access to its resources. Most of the accounts of project workers and farmers also indicate that these processes leave the growing popular aspirations generated by the government's discourse on development and economic growth mostly unattained.

Thus the making of the Beles Project highlights a main contradiction in the developmental state at work: on the one side, the territorialisation process related to the large-scale investments in irrigation allows the state, and in particular the federal government, to enhance its capacities of territorial control, resource accumulation and surplus extraction. On the other side, these large-scale projects contribute to social and economic complexity with increasing spatial mobility and social stratification that render a significant portion of the population – and their aspirations in terms of economic development – even more invisible and less intelligible to the state and its rigid structures of *encadrement*. Ultimately, the lack of recognition of people's claims and rights may jeopardize the political legitimacy and the capacity to control territories and resources of a developmental state that is still described as omnipresent and omnipotent.

Notes

1 This case study is part of a broader research project, "Accounting for Nile Waters", funded by the CGIAR Water, Land and Ecosystems program, combining the technical study of water productivity through remote sensing measurements with socio-political analysis on water, land and labour reallocations in the context of large-scale irrigation schemes in the Eastern Nile basin.

The work of Emanuele Fantini has been supported by the People Programme (Marie Curie Actions) of the European Union's Seventh Framework Programme (FP7/2007–2013) under REA grant agreement n. PCOFUND-GA-2013-606838.

We wish to acknowledge: Wondimu Wonde, Melisie Tolossa and Hayleyesusse Tameru for their contribution in collecting data between September 2015 and January 2016; Margreet Zwarteveen and Sabine Planel for their insightful comments on the previous versions of this chapter.

2 See for instance Meles Zenawi speech during the 13th Annual Pastoralists' Day celebrations in Jinka, South Omo, on 25 January 2011. www.mursi.org/pdf/Meles%20Jinka%20speech.pdf (latest retrieved 23 August 2017)

3 Data obtained from interviews to the general managers of Beles Project, AWWCE and ADWSE, December 2015.

4 "The Delayed Tana Beles Sugar factory costs Ethiopia 7 billion birr", *Welkessa*, July 2017. www.welkessa.com/the-delayed-tana-beles-sugar-factory-costs-ethiopia-7-billion-birr_6c3425b76.html (retrieved 25 August 25 2017).

5 "Sugar corporation in the red", *The Reporter*, 14 May 2016. http://archiveenglish. thereporterethiopia.com/content/sugar-corp-red (retrieved 4 July 2017).

6 Ecofin Agency 2016, *Ethiopia: Turkish firm Bedisa to acquire 75% stake in Beles sugar project*, www.ecofinagency.com/agriculture/2411-35931-ethiopia-turkish-firm-bedisa-to-acquire-75-stake-in-beles-sugar-project

7 Amba, Amhara Professionals Union 2016, "Stop BEDiSA Group's land grab of Tana Beles-Amhara region", *Ethiopia*. www.change.org/p/stop-bedisa-group-s-land-grab-of-tana-beles-amhara-region-ethiopia, 422 supporters as 16 June 2017.

8 Phone interviews with informant, Beles Project employee, June 2016.

9 http://ethiopiansugar.com/index.php/en/projects/belles-sugar-development-project (retrieved July 2017).

10 Interview with ESC irrigation engineer (M), 14 October 2015.

11 Interview with ESC plantation expert (M), October 2015.

12 Interview with ESC irrigation engineer (M), 20 October 2015.
13 Interview with ESC irrigation engineer (M), 13 October 2015.
14 Interview with ESC agronomist (M), 20 October 2015.
15 Interviews with Beles Sugar Project, AWWCE and ADWSE general managers, December 2015.
16 The lowest local authority in the Ethiopian institutional system, corresponding to a village.
17 Interview with daily labourer (M), 19 October 2015.
18 Interview with daily labourer (M), 14 October 2015.
19 Interview with daily labourer (F), 20 October 2015.
20 Interview with daily labourer (M), 14 October 2015.
21 Interview with daily labourer (M), 19 October 2015.
22 Interview with daily labourer (F), 12 February 2016.
23 According to ESC official data 4278 households for a total of 19,119 people have to be resettled in the Awi zone (Jawi *woreda*) and Metekel zone (Pawi and Dangur *woreda*).
24 Interview with resettled farmer (M), 17 October 2015.
25 Interview, Jawi Woreda agriculture office, head of food security, 20 October 2015.

References

Berhane-Selassie, T., 2009. Prestige is the argument: ideas of power and development aid in Ethiopia. In *Proceedings of the 16th International Conference of Ethiopian Studies* (Vol. 1), p. 4.

Berman, B. and Lonsdale, J., 1992. *Unhappy valley: conflict in Kenya & Africa* (Vol. 1). Athens: Ohio University Press.

Clapham, C., 2002. *Controlling space in Ethiopia.* In Donham, D., James, W., Kurimoto, E. and Triulzi, A. (eds.), *Remapping Ethiopia: socialism and after.* James Currey, Oxford, pp. 9–30.

Clark, A.C., Ratsey, J. and Wood, R., 2013. Feasibility studies for irrigation development in Ethiopia. *Proceedings of the Institution of Civil Engineers – Water Management, 166*(5), 219–230.

Dieci, P. and Viezzoli, C. (eds.), 1992. *Resettlement and rural development in Ethiopia: social and economic research, training and technical assistance in the Beles Valley.* Franco Angeli, Milan.

Fantini, E. and Puddu, L., 2016. Ethiopia and international aid: development between high modernism and exceptional measures. In *Aid and authoritarianism in Africa: development without democracy*, pp. 91–118. London: Zed Books.

FDRE (Federal Democratic Republic of Ethiopia), 2011. *Ethiopia's climate-resilient green economy strategy.* Addis Ababa (www.adaptation-undp.org/sites/default/files/downloads/ethiopia_climate_resilient_green_economy_strategy.pdf, latest retrieved on August 23rd, 2017).

Ficquet, E. and Feyissa, D., 2015. Ethiopians in the twenty-first century: the structure and transformation of the population. In Prunier, G. and Ficquet, É. (eds.), *Understanding contemporary Ethiopia: monarchy, revolution and the legacy of Meles Zenawi.* Oxford University Press, Oxford.

Fourie, E., 2015. China's example for Meles' Ethiopia: when development 'models' land. *The Journal of Modern African Studies, 53*(3), pp. 289–316.

Hagmann, T. and Péclard, D., 2010. Negotiating statehood: dynamics of power and domination in Africa. *Development and change, 41*(4), pp. 539–562.

Kamski, B., 2016. The Kuraz Sugar Development Project (KSDP) in Ethiopia: between 'sweet visions' and mounting challenges. *Journal of Eastern African Studies, 10*(3), pp. 568–580.

Lavers, T., 2016. Agricultural investment in Ethiopia: undermining national sovereignty or tool for state building? *Development and Change, 47*(5), pp. 1078–1101.

Lefort, R., 2007. Powers – mengist – and peasants in rural Ethiopia: the May 2005 elections. *The Journal of Modern African Studies, 45*(2), pp. 253–273.

Lefort, R., 2012. Free market economy, 'developmental state' and party-state hegemony in Ethiopia: the case of the 'model farmers'. *The Journal of Modern African Studies*, 50(4), pp. 681–706.

Lefort, R., 2015. Armanya: l'oignon, le masho, la bonne, le 'koulak' et le prolétaire. *EchoGéo, 31*.

Lund, C., 2016. Rule and rupture: state formation through the production of property and citizenship. *Development and Change*, 47(6), pp. 1199–1228.

Markakis, J., 2011. *Ethiopia: the last two frontiers*. Oxford: James Currey.

Moreda, T., 2017. Large-scale land acquisitions, state authority and indigenous local communities: insights from Ethiopia. *Third World Quarterly*, 38(3), pp. 698–716.

Mosley, J. and Watson, E.E., 2016. Frontier transformations: development visions, spaces and processes in Northern Kenya and Southern Ethiopia. *Journal of Eastern African Studies*, 10(3), pp. 452–475.

Pankhurst, A. and Piguet, F., 2009. *Moving people in Ethiopia: development, displacement & the state*. Eastern Africa Series. Oxford: James Currey.

Rahmato, D., 2011. *Land to investors: large-scale land transfers in Ethiopia*. Forum for Social Studies, Addis Ababa.

Scott, J.C., 1998. *Seeing like a state: how certain schemes to improve the human condition have failed*. New Haven and London, Yale University Press.

Villanucci, A. and Fantini, E., 2016. Santé publique, participation communautaire et mobilisation politique en Éthiopie: la Women's Development Army. *Politique africaine, 2*, pp. 77–99.

Wolde Selassie, A., 2009a. Identity, encroachment, and ethnic relations: the Gumuz and their neighbours in Northwestern Ethiopia. In Schlee, G. and Watson, E. (eds.), *Changing identities and alliances in North-East Africa* (Vol. 1), Oxford and New York: Berghahn Books.

Wolde Selassie, A., 2009b. Social impact of resettlement in the Beles valley. In Pankhurst, A. and Piguet, F. (eds.), *Moving people in Ethiopia: development, displacement & the state*. Eastern Africa Series. James Currey, Oxford.

World Bank, 2016. *5th Ethiopia economic update: why so idle? – wages and employment in a crowded labor market : draft for public launch*. World Bank Group, Washington, DC.

Zenawi, M., 2012. State and markets: neoliberal limitations and the case for a developmental state. In Noman, A., Botchwey, K., Stein, H. and Stiglitz, J. (eds.), *Good growth and governance in Africa: rethinking development strategies*. Oxford University Press, Oxford.

6 Water nationalism in Egypt

State-building, nation-making and Nile hydropolitics

Ramy Hanna and Jeremy Allouche

Introduction

The science of politics that characterizes IR-related hydropolitical theories might be too rigid to understand the various historical trajectories, diversity, and pathways of water-related interstate relations around the world. These perspectives provide little understanding on the social construction of norms and beliefs at the individual, societal, and transboundary levels. Social constructivism has barely been applied to hydropolitics (the only exception being Julien, 2012, at a theoretical level and a few studies on water securitization – see in particular Fischhendler, 2015). This book chapter will use this lens by focusing on the nation and the state as social constructs. This also means that the level of analysis will be on the interrelation between domestic politics and international relations, broadening out the analysis of insights on inter-state conflict and cooperation to multiple and interconnected scales (Menga, 2016). As most disputes occur following a unilateral action by a riparian to divert water or to build a dam, there is an urgent need to better understand the domestic politics leading to this particular decision.

The chapter presents an alternative analysis from the dominant hydropolitics literature. The chapter will look at 'water nationalism', a concept used to combine nation-making and state-building processes in defining domestic water politics and international transboundary water relations (see Allouche, 2005). Water nationalism, as a concept, will help to analyse the underlying dynamics of transboundary water conflicts by providing a major link between domestic policies, unilateralism, and international water politics. While nation-making and state-building have been put forward as important factors in explaining international conflicts, this type of analysis has not been applied to transboundary water politics. This concept is particularly useful in the context of Egypt where water was instrumental to broader colonial interests and post-colonial nationalistic movements (such as Nasserism) and a symbol of national pride and unity. It is also relevant to the Egyptian political economy where water is a central component to national development plans and the private sector's socio-economic interests.

Our overall argument is that nation-making and state-building processes may create tensions with the logic of sharing transboundary water resources.

We will demonstrate this conceptually, discursively, and empirically by showing how water nationalism informs the hydraulic mission of the modernizing and subsequently entrepreneurial state.

The presented case study examines the Nile water landscape in Egypt as a symbolic and imaginary construction of the nation state's geographical identity, which has important consequences on transboundary water relations. The 'hydraulic mission' of the state is defined and founded as the idea and practice of mastering nature and controlling the flow of water (Allouche, 2010; Allan, 2002). For Egypt, controlling water and achieving its hydraulic mission have been a source of state power and a key element of modernization across different historical phases. The chapter shows how the water nationalism discourse in Egypt has been – and still is – informed by the hydraulic mission of the modernization of the state, and its associated state-building and nation-making processes. The chapter analyses how Egypt's hydraulic mission was further developed by the entrepreneurial state, through mega projects and growing engagement in large-scale investments by corporate actors and foreign investors.

The chapter is divided into four sections. The first part examines how in Egypt water encompasses the elements of territoriality, sovereignty, and national identity as cornerstones informing state-building and nation-making processes across different historical periods. The second part focuses on the domestic and transboundary dimensions of nation-making and state-building processes in light of the hydraulic mission of the modernization of the state, especially during the Ottoman and post-colonial periods. The third part examines the hydraulic mission of the entrepreneurial state in the neoliberal economy as an extension of the hydraulic mission of the twentieth century. This section also examines the political economy of land and water investments as an integral part of the national economy and their implications on transboundary water management. In this respect it is argued that political nationalism has been complemented with economic nationalism, whereby acquisition of land and water resources for economic benefit has been translated into state-capital alliances. Section four concludes the chapter.

The hydraulic civilization: national identity and the historical roots of water nationalism discourse in Egypt

For millennia, water in Egypt signified the Nile. The perception of water as a natural gift and an essential element of national identity with cultural, political, and socio-economic connotations is reflected in Egypt's state-building and nation-making processes throughout history. As perhaps the most famous irrigated agricultural society in world history, the Egyptian civilization has conceived the Nile as an essential element of its geographical identity. Ancient Egyptians referred to their territory as *kemet*, the black or arable land, thus distinguishing the cultivable portion of their area from the desert, which they called *deshret*, or the red soil (Tignor, 2010). The English name *Egypt* derives from the Ancient Greek *Aígyptos* (Αἴγυπτος), which signified the waters whose annual flooding ensured the fertility of an earth (Kalin, 2006). The peculiar mystique of the Egyptian environment has historically informed the creation of its own sociological category: the hydraulic civilization.

The view that Egypt has always been and therefore should again become a society with a bureaucratic irrigation grid emerged within the milieu of French engineers towards the end of the eighteenth century (1799–1801). Napoleon Bonaparte was the first to recognize that the Nile afforded the government of Egypt a source of unusual power (Ludwig and Lindsay, 1937 in Kalin, 2006). The hydraulic mission of the state further evolved under the reign of the Ottoman viceroy Mohamed Ali Pasha (1805–1840), by means of a centralized apparatus of state denoting key attempts to modernize Egypt, thus marking a turning point in the agricultural history of Egypt (Kalin, 2006). Land and water have also been the foundation of Egypt's agrarian rural economy. Egypt's post-colonial 'hydraulic mission', led by Nasser's national movement, is marked by key nationalist milestones including agrarian reform, the construction of the Aswan High Dam, and the launch of state-sponsored large-scale agricultural and land reclamation projects.

Egypt's hydraulic mission towards modernization is centred on the main belief that water is available to pursuit the state's dreams to conquer the desert and establish new communities away from the 'overpopulated' Nile Valley and its delta. Over the last half a century, Egypt's hydraulic mission has been driven by what David Sims labels as an "ecological-demographic narrative of crisis" (Sims, 2015). The national vision of land reclamation, horizontal expansion, and greening the desert has been consistently part of the state's development discourse since 1952 to present times. Discursively, during the 1970s–1990s, water nationalism in Egypt became a symbol of national economic development through state-sponsored large-scale reclamation projects and the national vision for 'greening the desert' and extending the Nile's watershed. For the state, a key determinant of Egyptian water security is its framed vision to achieve its "desert development dreams" (see Sims, 2015), which also represents a main element of national security for social, economic, and political reasons. For thousands of years, Egypt's hydraulic mission has been historically relying on its water share from the Nile. Water played an essential role in informing Egypt's national identity, thus shaping state-building and nation-making processes and the wider water nationalism discourse.

State modernization and water nationalism in Egypt

From mastering nature to developing hydrocracies

As indicated by Molle (2009: 328), "water resources development by the state was an emergent and, at times, intentional, political strategy for controlling space, water and people and an important part of everyday forms of state formation" (see also Wester, 2008; Swyngedouw, 2007; Wehr, 2004; Reisner, 1993; Worster, 1985). Yet, "the necessary technology for large-scale perennial irrigation was

84   *Ramy Hanna and Jeremy Allouche*

unavailable until the nineteenth century A.D. when the traditional, basin or paleotechnic system . . . began to come to an end" (Butzer, 1976 in Kalin, 2006: 39). Mohamed Ali's 'hydraulic mission' began the shift in Egyptian agriculture from basin to perennial irrigation (Tignor, 2010). His vision and ambition was for Egypt to become a major world cotton exporter. To achieve this objective, he needed to provide regular quantities of irrigation water to the countryside during the low Nile season. The project required the improvement of existing canals, digging new canals, as well as the construction of dams and weirs along the Nile, for the sheer purpose of raising the water levels in the Nile and the branching canals for the farmers to irrigate summer crops.

> Where perennial irrigation was introduced, two, even three, crops could be grown per year, instead of the single crop that had been traditional for millennia. . . . The result was a large increase in government revenues, which, then, supported the enlarged army, the educational missions, the hydraulic improvements, and much else.
>
> (Tignor, 2010)

During the last two decades of the nineteenth century the colonial engineers with Indian experience arrived in Egypt (Willcocks and Craig, 1913) and were keen to reproduce the same hydraulic experiment. The hydraulic mission towards mastering nature and controlling the flow of water (Swyngedouw, 1999) inspired engineers and the government from the late nineteenth century, during what sociologists have termed industrial modernity (Beck, 1995, 1992). Whether out of a need to increase food production, raise rural incomes, or strengthen state-building and the legitimacy of the state, governments embraced the 'hydraulic mission' and entrusted it to powerful hydrocracies (Molle et al., 2009).

During the 1950s and 1960s Egypt's Nasser positioned water as a symbol of nationalism. The completion of the Aswan High Dam clearly represented this epoch and ideology. Water, irrigation, and agriculture along with their associated infrastructure were key ingredients of national mega projects for agricultural development. National mega projects were instituted as essential elements of state modernization, manifesting political and economic power and also informing the water nationalism discourse in Egypt.

Following 1952, the Egyptian government established agrarian reforms in the Nile Valley and Delta. Nasser redistributed, nationalized, and sequestrated land among landless farmers. The objective of the plan was to reform the Egyptian countryside with the imposition of a minimum agricultural wage and limited land distribution (Adriansen, 2009). Land reform entailed distributing reclaimed desert land to landless peasants under the 'Five Feddan Scheme' (Johnson, 2004). This program was an extension of early efforts since 1948 when the government announced a plan to distribute reclaimed desert land to small-scale farmers, who were to be given small plots of land and 'hygienic houses' (Mitchell, 2002: 40). Since the Agricultural Land Reform Act of 1952 (Law 178), land and water have reflected a sign of national pride and social justice for small farmers, and was therefore the first act in a radical reshaping of post-revolution Egypt.

After the 1952 Nasser era, land redistribution was an important goal, but land reclamation[1] was also on the agenda. The government launched a large-scale land reclamation program that was a continuation of an initiative of the colonial-era rural social improvement program (Mitchell, 2002) in the face of endemic disease and hunger after more than a century of intensified production of raw materials and food commodities for export to Europe (Dixon, 2013). The history of reclamation post revolution in Egypt during the 1960s was characterized by the development of large state farms with thousands of workers, inspired and partly financed by the Soviet Union (Springborg, 1979). However, the subsequent failure of these projects led to various attempts to subdivide the land between the workers in the 1970s (Adriansen, 2009). This means that in the older reclaimed lands, many of the inhabitants are small farmers and former workers from the state farms (Meyer, 1998, 1995 in Adriansen, 2009). Between 1964 and 1975, Egypt's post-colonial 'hydraulic mission' extended and became known as the 'Early High Dam Period'. It spans the time from the first closure of the Nile to the first filling of Lake Nasser reservoir's capacity to carry 164 billion cubic meters of water. The construction of the High Dam brought significant increases in the nation's welfare due to the reliable supply of adequate water for irrigation, as well as municipal and industrial use (Postel, 1996; Smith, 1986). With British assistance during the colonial period and by its own efforts afterwards, Egypt was able to establish relative water independence comparatively early by way of 'resource capture' (Zeitoun and Warner, 2006: 449) following the construction of infrastructural projects (such as the Aswan High Dam in 1971). This contrasts with the subsequent 'hydraulic missions' of upstream riparian states, which were traditionally impeded by their relatively weaker financial positions (Parkes, 2013; Cascão, 2009).

During the 'Post High Dam Period' (1976–1990s), the government continued its plan to expand land reclamation with an estimated 3 million feddan.[2] The development of land reclamation schemes depended on U.S. technology, reflecting Egypt's shift from the socialist camp to the Western neoliberal power. These plans benefited from imported technologies supplemented by USAID financing, supported irrigation infrastructure, canal lining, and high-tech machinery, amongst other water resource and agricultural development programs. The demand for water continued to rise due to population growth and economic development, whereby available resources were becoming insufficient to meet the expected demand from competing users (Abu-Zeid, 2003). Economic considerations were also a priority issue for the state, whereby the economic value of water was reflected in the agricultural sector's contribution to the economy, representing 15–20% of Egypt's GDP while employing 30% of the total workforce. Responding to the challenge of water scarcity, the government pursued a two-fold policy of supply enhancement and demand management. On the supply side, the government planned to develop new water resources in cooperation with riparian countries, by rainfall harvesting and through limited desalination. Major water savings were expected from the adoption of demand management policies in the old land that encouraged the reuse of drainage water, the elimination of Nile water flows to the sea, and the irrigation improvement program. The state-centric development plans for land reclamation and greening the desert targeted the west of Delta and Al Nubareya, in addition to several donor-funded development projects that have been deployed over the last four decades including the World Bank New Land Development Project (1980–1991), the Mubarak Project for Developing and Serving the Land Allocated to Youth Graduates initiated in 1987, and the West Delta Irrigation Improvement Project in 2005.

During the second half of the twentieth century, there was a wide consensus that land reclamation schemes were potentially an alternative pathway to address youth unemployment and divert the traditional mindset of expecting a government job upon graduation. They also served the larger government vision of greening the desert – a big slogan throughout the 1980s, also revived in modern times under the '1.5 million feddan' project inaugurated in 2015. Others however, including Mitchell (1995) and Bush (2007, 2002), have contested this dominant discourse of state-led development, and view these land reclamation schemes as a way to avoid the reform and redistribution of the old Nile Valley and Delta lands. They argue that these policies should be understood as part of the wider political maneuvering of international partners such as USAID to increase the country's reliance on animal production at the expense of plant production, leading to growing U.S. wheat imports (Mitchell, 1995).

In the new national imaginary, land reclamation was made inseparable from the Nile Valley and Delta. The post-colonial hydraulic mission of the Egyptian state has been driven by a twin legitimating discourse of too many people in too little land in the Nile Valley and Delta, and of remaking citizens in the *tabula rasa* of the desert (Sowers, 2011). The desert was proposed to be carved out as a *terra nullius*, an undeveloped space to be developed; it became a tool of nation-making. New citizens would build modern farms in the blank landscape of the desert (Sowers, 2011). Large-scale state-sponsored schemes such as Al Nubareya and Al Salheya were also symbols of modernization and the larger vision of the state to green the desert and expand Egypt's population beyond the narrow valley of the Nile and its overcrowded delta downstream.

Throughout the last half a century and more recently during the first two decades of the twenty-first century Egypt's 'national' hydraulic mission towards modernization and the state's framing of water security were largely driven by a narrative of crisis. This narrative had the twin objectives of 'horizontal expansion' and 'demographic redistribution' to expand the Nile's watershed away from the overcrowded Nile Valley and its delta, through land reclamation and by establishing large-scale agricultural projects for greening the desert, adopting a 'mega' project approach.

National hydraulic mission and transboundary hydropolitics

The transboundary hydropolitics of the Nile River basin can be argued to have started taking shape during the nineteenth century. The early modernization efforts of the Egyptian state, led by Mohamed Ali, were a main driver for the control of water resources beyond Egypt's national territory, which in turn informed and shaped the early water nationalism discourse in Egypt. This transboundary interest was manifested in Mohamed Ali's orders to send a campaign to Sudan under the leadership of Ismail Pacha in 1820–1821. Later on, the colonial economic and geo-political interests of the British Empire in Egypt and Africa further took the Egyptian domestic interests in transboundary water resources to a higher level by establishing water rights treaties (e.g.1902[3] and 1929).[4] Whether for national or colonial interests, the realization that water is an essential element for Egypt's social, economic, as well as its strategic political interests was realized as early as the nineteenth century. These milestones mark the historical importance of water on the domestic and transboundary levels, which in turn informed the water nationalism discourse.

Egypt's post-colonial domestic hydraulic mission during the High Dam period (1964–1975) marks a peak of Egyptian nationalism. Turton (2000) indicates that this is a turbulent era where the dynamics of the Cold War become inter-linked with national politics at the basin level. It is characterized by zealous patriotism that sees the full effect of the 'High Dam Covenant' on political decision-making. The involvement of the Soviets in the construction of the High Aswan Dam was not welcomed in the West, and led to Cold War recriminations about the hydropolitics of the region (Turton, 2000). During most of the second half of the twentieth century, the domestic interests of Egypt and Sudan shaped much of the Nile's hydropolitics on the transboundary level. The Nile's hydropolitics were mainly driven by the quest for post-colonial state modernization, which in turn informed the water nationalism discourse. On the transboundary level, for a long period of time extending between the 1959 agreement up to the launch of donor-supported Nile Basin Initiative (NBI) in 1999, the status quo in Nile hydropolitics prevailed, thus leading to contestations by upstream countries seeking to achieve their own domestic modernization efforts.

The hydraulic mission of the entrepreneurial state

We extend our view of the role of water in state-building and nation-making processes by examining the hydraulic mission of the entrepreneurial state. The idea of the entrepreneurial state may appear as a contradiction in terms. For neo-liberals such as Friedman (2009), grand visions and innovations typically came from "pioneers, hackers, inventors, and entrepreneurs," not from the lumbering actions of the bureaucratic state (see also Swedberg, 2015; Isaacson, 2014). This idea has been increasingly called into question. This section shows how Egypt's national development paradigm shifted from modernization to an entrepreneurial state with respect to its hydraulic mission domestically and its relation with transboundary hydropolitics.

Domestic determinants of the entrepreneurial state

Mazzucato (2015) in her book *The Entrepreneurial State* argues that the state plays a pivotal role in shaping the modern economy, industrializing entire economies and supporting the private sector in a myriad of ways (from publicly funded research activities, small business start-up grants, and tax credits, amongst other measures). The modern state does not only nudge the economy, it actually pushes it forward through bold and innovative measures (Mazzucato, 2015). This idea of the state's role in shaping and creating markets is more in line with the work of Karl Polanyi (2001 [1944]), who emphasized how the capitalist 'market' has from the start been heavily shaped by state actions. State entrepreneurship in this respect is a means through which governments in developing countries forge and maintain a political alliance with the corporate sector to achieve their hydraulic mission. In Egypt, this 'state-capital' alliance couples the political and domestic development visions of greening the desert, achieving food sovereignty, and self-reliance of key strategic crops. Overall, there has been little systematic inquiry into the processes through which state agencies take on a direct entrepreneurial role by supporting upstream value-chain investments in land reclamation and mega infrastructure projects, thus paving the way for the private sector to further develop large-scale agricultural projects.

The hydraulic mission of the 'entrepreneurial state' can be seen as part of "frontier making or the expansion of socio-ecological spaces for capital accumulation" (Dixon, 2013). In this respect, this process of frontier-making and expansion into the desert was driven by Egyptian state ecological-demographic crisis narratives. It has also led to a central condition of heightened capital accumulation[5] in agriculture and food in Egypt throughout the economic liberalization program. Since May 1991, the state under President Mubarak (1981–2011) turned to encouraging private sector participation in large-scale agricultural investments. The Egyptian government continued to pursue its national plan for land reclamation, but this time with direct participation of the private sector. These policies displaced the political and economic policies of the former socialist regime under Nasser (1952–1970). By doing so, the state shifted from being the actual developer, operator, and manager of agricultural projects to being an investor in infrastructure for large-scale projects. The evolution of its role in achieving its hydraulic mission comes in line with the larger vision, which entails state-sponsored investments in desert infrastructure, and risk-taking to establish new communities outside the Nile Valley. Public and private capital flowed into technology-intensive mega hydraulic and land reclamation projects such as Toshka, Oweinat, and the recently established 1.5 million-feddan project launched in 2015. As such, beyond the private farmlands depending on underground water to export high-value crops to European markets, investing in establishing sustainable communities outside of the Nile Valley has been usually considered a risky task for private investors. In this respect, especially in the context of Egypt, the state has acted as a force for change, not only 'de-risking' the economic landscape for risk-averse private actors, but also leading the way with an ambitious project (Mazzucato, 2015).

Both Toshka and Sharq Al-Owainat inaugurated in 1997 and 2001 respectively – also named the 'South Valley Project' – are good examples of mega projects sponsored by the entrepreneurial state. These projects aim to create a new valley for Egypt's Nile River to be situated in the western desert, about 900 km south of Cairo and 100 km from the southern borders with Sudan. The initial intention was to allow 6 million people to live there by 2017, in addition to creating 10,000 job opportunities. The project aimed at reclaiming 540,000 feddan for agriculture depending on water resources from Lake Nasser, supplemented by underground water from the Great Nubian Sand Aquifer. The key government narrative behind these projects is as follows: firstly, the projects are seen as a way to address the demographic issue in Egypt; the projects will enable future generations to escape the narrow strip of the Nile Valley and its overcrowded delta. The second narrative was that these projects are national symbols of modern economic Egyptian development that will ensure food and national security.

These 'national' projects are in some ways marketed as a twenty-first century national project reflecting the initiative of the entrepreneurial state and its hydraulic mission. Land reclamation schemes are central to this mission aiming to expand agriculture beyond the natural extensions of the old Nile Valley into Egypt's abundant desert land, with an economic logic to attract investors and capital-driven investments. They depend on private investors to inject the necessary capital and know-how to develop large-scale profitable agricultural projects using water-saving modern irrigation techniques. As such, securing water resources for these large-scale private investments has become a national priority to further advance the hydraulic mission of the entrepreneurial state, which discursively aims at redistributing the population, greening the desert, attracting foreign investments, and securing the maximum financial returns from land and water resources. In turn, given that securing water for corporate investments is an essential factor in the process of state-building, it can be also claimed as an additional factor defining economic and water nationalism in Egypt.

The food and fuel crises of 2008 and 2011 also boosted the plans of the entrepreneurial state, whereby commodity prices witnessed historical hikes, thus leading global demand for land and water resources to grow significantly (Allouche et al., 2015). Large-scale agricultural projects were encouraged to set the model of state-of-the art commercial production, advanced agricultural machinery, and a source of foreign currency from land leasing. They were also viewed as a source of inflow of investments in the agriculture sector, ultimately leading to food production and employment creation, thus fulfilling the state's vision of horizontal expansion and establishing sustainable communities in the desert. The state-capital alliance towards large-scale land acquisitions (LSLA) and agricultural projects has been therefore a key avenue for 'frontier-making' within the borders of what is presently modern-day Egypt. In today's global economy and its connectedness, 'national frontier-making' here is connected to frontiers regionally, especially as it relates to key strategic issues such as water and food security.

Given this understanding of the Nile water symbolism in Egypt and its associated hydraulic mission of the entrepreneurial state, it is important to discuss how this positioning of Egyptian national hydropolitics has influenced transboundary water resources in the Nile basin.

The entrepreneurial state and transboundary water management

At the transboundary level, during the first decade of the twenty-first century the hydraulic mission of the entrepreneurial state collided with two key events reflecting contestations around Nile hydropolitics. The first is the contestations by upstream Nile River basin countries about the international water treaties of 1929 and 1959, and the limitations of the donor-supported cooperation mechanism by the Nile Basin Initiative (NBI) initiated since 1999. These events led some upstream countries to support the Cooperative Framework Agreement (CFA) declared on May 14, 2010, attempting to replace the historical Nile treaties. This resulted in hydropolitical tensions, leading to the Egyptian state to withdrawal from the NBI. The second is Ethiopia's unilateral launch of the Grand Ethiopian Renaissance Dam (GERD) in 2011, which was viewed as a direct threat to Egypt's transboundary interests. These events were interpreted by Egypt to impose some limitations on the amount of water to reach Lake Nasser. Also, they were largely perceived as imposing a direct risk to farmers' livelihood downstream, as well as a limitation over Egypt's hydraulic mission as an entrepreneurial state, seeking to attract local and international agricultural investments, which constitute a major element for its economic growth. While Egypt needed to preserve its share of the Nile's water to further develop its hydraulic mission during the twenty-first century, these events heightened the water nationalism discourse, not only in Egypt, but also in other Nile basin countries, notably Ethiopia.

Furthermore, during the last decade, and in light of the prevailing neoliberal global economy, new actors have entered the hydropolitics scene through large-scale capital investments and large-scale land acquisitions in the agricultural sector. Specifically in Africa, large-scale land acquisitions have been taking place in unprecedented volume and intensity in recent years (Odusola, 2014). A 'scramble' for Nile basin riparian agricultural land, by governments and private companies, for the production and export of cereals for food or use as biofuel, came to the fore. Evidence indicates that new capitalist players, private actors, and international investors aim to develop those lands with industrial agricultural methods (Clapp and Helleiner, 2012: 499). Different media and academic reports have attributed this phenomenon to the food and energy crises of 2008 and 2011, which witnessed the highest peak of the global commodity index since 1845 (Allan, 2013). These dynamics of the global economy led to a renewed interest in large-scale land acquisitions and agricultural – hence water – investments by the private sector globally. Non-state actors in the form of sovereign wealth funds, private equity funds, international agricultural companies, as well as other private and financial institutions (FIs) investing abroad are motivated by water and food security objectives, as well as profitability and financial return.

This increasing competition towards controlling natural resources was labelled "land and water grabs", defined as "the large-scale acquisition of land in developing, economically and politically weak countries through FDI by powerful, developed, economically robust, but water-stressed, nations" (Sebastian and Warner, 2013: 2). Evidence suggests that many African countries are "motivated by potential revenues from water fees and the prospect of improved agricultural productivity, signing away water rights for decades to large investors" (IIED, 2011). Many of the capitalist players and non-state actors originate from water-scarce countries and are often supported by their home governments to invest abroad in order to secure a country's needs of certain strategic crops, which cannot be homegrown due to lack and scarcity of water resources (Hanna, 2016). Others invest for the sheer objective of cultivating high-value exportable crops and agricultural commodities tradable in the global market, denoting a form of the financialization of natural resources (Scoones et al., 2015).

These projects are part and parcel of the hydraulic mission of the entrepreneurial state and their quest to attract investments using their available land and water resources. As such, 'land and water' resources in transboundary river basins such as the Nile in Africa have become prime target destinations for these investments and their associated large-scale land acquisitions and water resource use. Egypt, Ethiopia, and Sudan are particularly attractive countries, due to their available land and water resources. While these investments are directly linked to the hydraulic mission of the entrepreneurial state in Egypt and other Nile basin countries, they may represent a threat to the sustainable use of water resources, and will impose additional pressure on already strained water supplies upstream. Simultaneously, financial and technical assistance offered by external actors (as part of an overall strategy to facilitate market access) has also allowed the previously unfeasible construction of upstream dams and the expansion of irrigation for agriculture, which further complicates the issue of Egyptian water security. Achieving its hydraulic mission as an entrepreneurial state and future expansions may be therefore negatively affected by upstream investments, of both the state and the corporate actors.

Therefore, transboundary hydropolitics influence and are influenced by the potential competition over water resources amongst different upstream and downstream countries that seek to achieve their own *national* hydraulic missions on the domestic scale. Given that water nationalism is informed by the hydraulic mission of the state as a key element in state-building and nation-making processes, this competition may result in growing tensions over transboundary water allocations and use. In the case of Egypt, the attempts of the modern state have clearly shaped the Nile's historical hydropolitics, while in contemporary times, the hydraulic mission of the entrepreneurial state is colliding with upstream developments and a changing hydropolitical landscape of the Nile basin given the global political economy. As such, domestic and transboundary hydropolitics are constituent of each other and result in creating tensions between national visions and international politics at the geographical scale of the transboundary river basin.

Conclusion

This chapter introduced the concept of 'water nationalism' and focused on the particular history of Egyptian state modernization within the larger context of the Nile River basin hydropolitics. The analysis examined the historical role of water in the nation-making and state-building processes at the national scale based on the historical and contemporary hydropolitics of Egypt during colonial, post-colonial, and neoliberal eras within the larger context of the Nile River basin. It showed how water bodies, infrastructures, and landscapes became part of the broader national identity and 'hydraulic mission' of the state as an essential element of state-building and nation-making.

By examining the state-society-nature interactions on domestic and transboundary scales, the contribution moved away from the environmental security and scarcity lens, as well as the traditional conflict-cooperation narrative in hydropolitical analysis, towards a broader state-society understanding of hydropolitics. The chapter discussed how domestic water politics inform the water nationalism discourse in Egypt, whereby the hydraulic mission of the state, and its modernization narratives, influence and are influenced by the Nile basin hydropolitics. This two-way relationship reflects how the domestic hydraulic mission and transboundary hydropolitics are constitutive of each other.

On the one hand, water nationalism evolved in Egypt on the domestic and national scales across different historical phases, situated within state-building and nation-making processes, which we consider informed by (i) national identity including territorial sovereignty and geographical identity, (ii) the hydraulic mission of the modernizing state, and (iii) the hydraulic mission of the entrepreneurial state. In this respect, the Nile River as the only source of renewable water resources in Egypt has been a key element of national security and economic development since ancient periods to contemporary times. These factors shaped Egypt's water nationalism discourse especially during post-colonial times and through the rise of political nationalism, as well as the contemporary neoliberal economy and economic nationalism, not only in Egypt but in other upstream riparian countries as well.

On the other hand, Nile transboundary hydropolitics influence(d) and are influenced by this national framing of water nationalism both historically, as well as in contemporary times, within the larger global political economy. Historically rulers and colonizers have realized the importance of water resources for the fulfilment of Egypt's hydraulic mission and its political and economic prosperity, thus resulting in their seeking to secure upstream resources through Ottoman explorations and legal treaties during colonial and post-colonial periods. In contemporary times, upstream mega infrastructure projects such as dams (i.e. the GERD) and large-scale irrigation projects in Nile basin countries have increased hydropolitical tensions with Egypt downstream as they may influence its water share and consequently its hydraulic mission as an entrepreneurial state. Furthermore, given today's neoliberal economy, the growing competition over water resources from non-state and corporate actors for financial profit aggravates the issue of competition due to commercialization considerations. Thus, in light of today's neoliberal global economy, the evolving hydraulic mission of the entrepreneurial state, an extension of the hydraulic mission of the twentieth century, implies that political nationalism has been complemented with economic nationalism.

The case of water nationalism in Egypt was selected due to its historical and contemporary relevance. It provides an example linking international hydropolitics with domestic nation-making and state-building processes, as well as the larger hydraulic mission of the state, which evolved throughout colonial, post-colonial, and contemporary periods. Water nationalism also explains the influence of the various cultural, symbolic, socio-economic, and political economic manifestations of state-building and nation-making historical and contemporary processes over transboundary water politics in the Nile basin.

Notes

1 For the non-Egyptian reader, there is a need to clarify the meaning of reclamation as it is used in Egypt denoting farming the desert land. This meaning is different than the traditional sense of the term, mainly applying to lands lost through poor drainage, salinity, and other water management–related practices located in or near the Old Valley, which were reclaimed before 1982 (Zalla et al., 1998).
2 The unit of area measurement in Egypt. One feddan is made up of 24 qirat and is equivalent to 0.42 hectares or 1.04 acres.
3 The first Nile agreement to be signed was in 1902 between Ethiopia and Britain on behalf of Egypt. The Addis Ababa agreement marks the "principle of non-interference with the flow of Blue Nile" and Ethiopia's first agreement to seek prior consent before initiating works affecting the flow of the Blue Nile or Sobat.
4 The Nile Water Agreement signed between Britain – representing the Sudan, Kenya, Tanganyika, and Uganda – and the Egyptian governments; the Belgium colonies were not signatories to this (Zaire, Rwanda, and Burundi). Through the 1929 agreements, Egypt received 48 cubic km of the flow and full access to the spring flood, while the Sudan only has a claim to 4 cubic km of the river flow (Haftendorn, 2000: 58).
5 Other economic liberalization programs took place during the nineteenth century and first half of the twentieth century before applying the socialist economic system since Nasser in 1952.

References

Abu-Zeid, K., 2003. *Potential for water savings in the Mediterranean region*. World Water Forum, Kyoto, 16–23 March 2003. CIHEAM-IAMB.

Adriansen, H.K., 2009. Land reclamation in Egypt: A study of life in the new lands. *Geoforum*, *40*(4), 664–674.

Allan, J.A., 1999. The Nile basin: Evolving approaches to Nile waters management. *SOAS Occasional Paper*, *20*, 1–11.

Allan, J.A., 2002. *The Middle East water question: Hydropolitics and the global economy*. London and New York: I.B. Tauris.

Allan, J.A. (Ed.), 2013. *Handbook of land and water grabs in Africa: Foreign direct investment and food and water security*. New York: Routledge.

Allouche, J., 2005. *Water nationalism: An explanation of the past and present conflicts in Central Asia, the Middle East and the Indian Subcontinent?* PhD thesis, Graduate Institute of International Studies.

Allouche, J., 2010. The multi-level governance of water and state building processes: A longue durée perspective. In Wegerich, K. and Warner, J. (Eds), *The politics of water: A survey*, pp. 45–67. London, New York: Routledge.

Allouche, J., Middleton, C. and Gyawali, D., 2015. Technical veil, hidden politics: Interrogating the power linkages behind the nexus. *Water Alternatives*, 8(1), 610–626.

Beck, U., 1992. *Risk society: Towards a new modernity*. London: Sage.

Beck, U., 1995. *Ecological politics in an age of risk*. Cambridge: Polity Press.

Bush, R., 2002. Land reform and counter-revolution. In Bush, R. (Ed.), *Counter-revolution in Egypt's countryside: Land and farmers in the era of economic reform*. Malaysia: Zed Books, 1–31.

Bush, R., 2007. Politics, power and poverty: Twenty years of agricultural reform and market liberalisation in Egypt. *Third World Quarterly*, 28(8), 1599–1615.

Butzer, K.W., 1976. Early hydraulic civilization in Egypt: A study in cultural ecology. *Prehistoric Archaeology and Ecology (USA)*.

Cascão, A.E., 2009. Changing power relations in the Nile river basin: Unilateralism vs. cooperation? *Water Alternatives*, 2(2), 245.

Clapp, J., and Helleiner, E., 2012. Troubled futures? The global food crisis and the politics of agricultural derivatives regulation. *Review of International Political Economy*, 19(2), 181–207.

Dixon, M., McMichael, Philip, Makki, Fouad, Mitchell, Timothy, & Santiago-Irizarri, Vilma. (2013). *The Making of the Corporate Agri-food System in Egypt*, ProQuest Dissertations and Theses.

Fischhendler, I., 2015. The securitization of water discourse: Theoretical foundations, research gaps and objectives of the special issue. *International Environmental Agreements: Politics, Law and Economics*, 15(3), 245–255.

Friedman, M., 2009. *Capitalism and freedom*. Chicago: University of Chicago Press.

Haftendorn, H., 2000. Water and international conflict. *Third World Quarterly*, 21(1), 51–68.

Hanna, R.L., 2016. Transboundary water resources and the political economy of large-scale land investments in the Nile. In Sandstrom, E., Jagerskog, A. and Oestigaard, T. (Eds.), *Book chapter in land and hydropolitics in the Nile River basin: Challenges and new investments*. London: Routledge.

IIED, 2011. *Are land deals driving 'water grabs'? Briefing: The global land rush*. London: International Institute for Environment and Development (IIED). http://pubs.iied.org/17102IIED, accessed September 2014.

Isaacson, W., 2014. *The innovators: How a group of inventors, hackers, geniuses and geeks created the digital revolution*. New York: Simon and Schuster.

Johnson, A.J., 2004. *Restructuring rural Egypt: Ahmed Hussein and the history of Egyptian development*. Cairo: The American University in Cairo Press.

Julien, F., 2012. Hydropolitics is what societies make of it (or why we need a constructivist approach to the geopolitics of water). *International Journal of Sustainable Society*, 4(1–2), 45–71.

Kalin, M., 2006. *Hidden pharaohs: Egypt, engineers and the modern hydraulic*. Master of Philosophy dissertation, University of Oxford, Oxford.

Ludwig, E. and Lindsay, M.H., 1937. *The Nile: The life-story of a river*. New York: The Viking Press, 19–22.

Mazzucato, M., 2015. *The green entrepreneurial state*. SPRU Working Paper Series (October), University of Sussex.

Menga, F., 2016. Domestic and international dimensions of trans-boundary water politics. *Water Alternatives*, 9(3), 704–723.

Meyer, G., 1995. Liberalisierung und privatisierung der ägyptischen landwirtschaft. *Erdkunde* 49, 17–31.

Meyer, G., 1998. Economic changes in the newly reclaimed lands: from state farms to small holdings and private agricultural enterprises. *Directions of change in rural Egypt*, pp. 334–356.

Mitchell, T., 1995. The object of development: America's Egypt. In Crush, J. (Ed.), *Power of development*. London: Routledge, 129–157.

Mitchell, T., 2002. *Rule of experts: Egypt, techno-politics, modernity.* Berkeley: University of California Press.

Molle, F., Mollinga, P.P. and Wester, P., 2009. Hydraulic bureaucracies and the hydraulic mission: Flows of water, flows of power. *Water Alternatives, 2*(3), 328–349.

Odusola, A.F., 2014. *Land grab in Africa: A review of emerging issues and implications for policy options.* Working Paper No. 124, International Policy Centre for Inclusive Growth.

Parkes, L., 2013. The politics of 'water scarcity' in the Nile basin: The case of Egypt. *Journal of Politics & International Studies, 9*(Summer), 433–480.

Polanyi, K., 2001 [1944]. *The great transformation: The political and economic origins of our times.* Boston: Beacon.

Postel, S., 1996. *Dividing the waters: Food security, ecosystem health, and the new politics of scarcity.* Washington, DC: Worldwatch Institute.

Reisner, M., 1993. *Cadillac desert: The American West and its disappearing water* (Revised and updated). New York: Penguin Books.

Scoones, I., Leach, M. and Newell, P. eds., 2015. *The politics of green transformations.* London: Routledge.

Sebastian, A.G. and Warner, J.F., 2013. Geopolitical drivers of foreign investment in African land and water resources. *African Identities*, 1–18.

Sims, D., 2015. *Egypt's desert dreams: Development or disaster?* Oxford: Oxford University Press.

Smith, S.E., 1986, October. General impact of Aswan high dam. *Journal of Water Resources Planning and Management* (New York), *112*(4), 551–562.

Sowers, J., 2011. Re-mapping the nation, critiquing the state: Environmental narratives and desert land reclamation in Egypt. In Diana K. Davis and Edmund Burke (Eds.), *Environmental imaginaries in the Middle East: History, policy, power, and practice*, 158–191. Athens: Ohio University Press

Springborg, R., 1979. Patrimonialism and policy making in Egypt: Nasser and Sadat and the tenure policy for reclaimed lands. *Middle Eastern Studies, 15*(1), 49–69.

Swedberg, R., 2015. The state, the market, and society. *Contexts, 14*(3), 58–59.

Swyngedouw, E., 1999. *Modernity and hybridity – The production of nature: Water an modernisation in Spain.* Paper presented to the SOAS Water Issues Study Group, University of London, 25 January 1999.

Swyngedouw, E., 2007. Technonatural revolutions: The scalar politics of Franco's hydro-social dream for Spain, 1939–1975. *Transactions of the Institute of British Geographers, 32*(1), 9–28.

Tignor, R., 2010. Chapter 1: The land and people. In *Egypt: A short history.* Princeton: Princeton University Press.

Turton, A.R., 2000. *A cryptic hydropolitical history of the Nile basin for students of hydropolitics.* Pretoria: Study Guide for Pretoria University.

Wehr, K., 2004. *America's fight over water: The environmental and political effects of large-scale water systems.* New York and London: Routledge.

Wester, P., 2008. *Shedding the waters: Institutional change and water control in the Lerma-Chapala basin, Mexico.* PhD thesis, Wageningen University, Wageningen, the Netherlands.

Willcocks, W. and Craig, J., 1913. *Egyptian irrigation*, 3rd ed., 2 vols. London and New York: E. & F. N. Spon Ltd; Spon & Chamberlain.

Worster, D., 1985. *Rivers of empire: Water, aridity, and the growth of the American West.* Oxford: Oxford University Press.

Zalla, T., Goetz, S., Holtzman, J., Young, R., Saad, A.H.Y., Fawzy, M.A. and Ismail, A.R., 1998. *Monitoring, verification, and evaluation unit: Agricultural policy reform program.* Cairo: Abt Associates. USAID.

Zeitoun, M. and Warner, J., 2006. Hydro-hegemony – A framework for analysis of trans-boundary water conflicts. *Water Policy, 8*(5), 435–460.

7 Troubled waters of hegemony

Consent and contestation in Turkey's hydropower landscapes[1]

Bengi Akbulut, Fikret Adaman and Murat Arsel

Introduction

If the idea of development has been central to realising the dream of elevating (or rather returning) Turkey to the status of a "great state", hydropower has been one of the key weapons of the state as it sought to transform its putatively backward society. For much of modern Turkish history, therefore, the state mobilised the nation's riverine resources against itself, to forcefully modernise the peasantry in a bid to make them jump, in former Prime Minister Özal's famous parlance, into a "new epoch" (one that Western civilizations had inhabited for some time). To the extent that there has been limited societal resistance to the state's increasingly grandiose hydropower projects, this has been partly a function of the Turkish peasantry's internalisation of the argument that it does indeed require dramatic transformation at the hands of a powerful state.

Yet the last decade has seen a dramatic reversal in the shape of widespread and vigorous social conflicts against hydroelectric power generation projects across the nation. What is surprising is not only the dramatic reversal itself, but also its target. Rather than questioning the large (and at times mega) dam and irrigation projects, the emerging resistance has focused on small-scale, diversion-type run-of-river hydropower plants, which re-route streams to a suitable height by covered channels from where they are dropped on turbines, that began to mushroom circa 2000 in the hills and mountains of Turkey. These plants are exclusively called *Hidro Elektrik Santralı (HES)* in Turkish, and will be referred to as HPPs (a direct translation) throughout the chapter.

It is this reversal that the chapter is focused on, by situating it within the broader historical dynamic between state and society through the deployment of the Gramscian concept of hegemony. We argue, in particular, that while dams as a form of hydro-technology could animate modernisation and development as collective interest on which the Turkish state's hegemonic project is built on, HPPs failed to do so and thus were, and still are, widely opposed. In doing so, we emphasise how the notions of mutuality and reciprocity mobilised by dams enabled a more effective constitution of the collective interest around them. We illustrate our argument via findings from an extensive field study conducted in Artvin, a province in the north-eastern corner of Anatolia. More specifically, our study focuses on the Ardanuç-Şavşat-Borçka triangle in the north of the province, where we conducted fieldwork intermittently between 2012 and 2014.[2]

From mega dams to HPPs: shifting technologies of hydropower in Turkey

Initiated in the late 1950s and gaining momentum especially in the 1960s and 1970s, building dams became Turkey's main hydraulic strategy, much like its developing country counterparts. The founding of the State Hydraulic Works (SHW) in 1953, responsible for the development and management of water resources, was a crucial step in this direction, followed by the construction of Turkey's first two major dams, Seyhan and Sarıyar, in 1956. Developing the hydro-energy potential of the country had by then been fully endorsed as a crucial step towards rapid, industry-oriented economic development, which was, in turn, seen as a *sine qua non* for modernisation by the state, whose ambitious pursuit of this goal raised the share of hydro-energy from 4% in 1950 to 30% in 1960 (Evren, 2014). The planning and construction of big dams accelerated significantly after the 1960s, and the gigantic Southeast Anatolian Project (SAP) deserves mention in particular. Comprising a comprehensive plan of regional development in agriculture, energy, transportation and infrastructure in addition to a massive scheme of hydro-infrastructures, the SAP covers about 75,358 square kilometres – almost 10% of Turkey's total surface area. The project envisioned the construction of 22 dams (19 of which are completed as of today), which would enable the irrigation of 1.8 million ha of land and the production of 27 billion kWh of energy annually[3] (Çarkoğlu and Eder, 2003, 2005; Harris, 2002; Özok-Gündoğan, 2005). Reflecting the state's profound commitment to dam building as a development strategy, the SAP has been depicted as a silver bullet for various regional and national ambitions, be it increased agricultural yields, the empowerment of women or the resolution of ethnic conflict.

At first operationalised through an institutional constellation of state organs, such as the State Planning Agency and the SHW, and international donors such as the World Bank, the planning, building and operation of hydro-infrastructures passed through several waves of liberalisation, starting in the early 1980s. Yet the construction of dams and associated monumental irrigation schemes remained a prominent practice under the orchestration and control of the state until the mid-2000s. The consolidation of energy market liberalisation during this period was perhaps the most significant marker of the latest neoliberal turn in Turkey. While certainly not unprecedented (Erensü, Evren and Aksu, 2016), the 2005 law that deregulated the electricity market enabled private producers to participate more extensively in energy generation projects and provided market and credit incentives to such investments (Adaman, Akbulut and Arsel, 2017a). Coupled with the systematic relaxation of environmental regulations, this wave of liberalisation not only lead to a boom in the energy sector but also significantly reshaped the country's energy topography.

Within the field of hydro-energy in particular, the early 2000s marked a shift both in the dominant form of hydropower investments and the actors undertaking their development. This period was characterised by the substantial increase in private sector involvement in hydropower, following the interventions that finally rendered the field of hydro-energy sufficiently attractive for investment. Concomitantly, direct state involvement in the construction, development and operation of hydropower plants fell significantly: the ratio of private to public in installed capacity in hydropower rose from 1/10 in 2003 to 3/2 in 2015, and the share of private investment in activated installed capacity of hydropower between 2003 and 2015 was 80% (Erensü, Evren and Aksu, 2016). That by the end of 2016 the share of private investment (by installed capacity) was 45% in existing hydropower plants, 64% in plants under construction and 94% in plants being planned clearly indicates that the state's role in hydro-energy is shrinking[4] (DSI, 2016).

This shift from state to private investment and operation was accompanied by a change in the dominant form of hydropower technology in energy production. The prevalence of dams and associated large hydroelectric plants was gradually eroded by the boom in HPPs in quantity as well as in scope, though not in terms of installed/production capacity. Of the production permits issued for hydropower between 2003 and 2016, 74% were for plants with an installed capacity less than 25 MW (and almost 50% were for plants with an installed capacity less than 10 MW), almost all of which are HPPs (www.enerjiatlasi.com/hidroelektrik/; Erensü, Evren and Aksu, 2016). Furthermore, the overwhelming majority of HPPs constructed in this period were private sector investments, which constitutes the point of overlap between the two shifts identified above in terms of hydropower investments and the actors undertaking their development.

The mushrooming of HPPs especially after the mid-2000s was met with an equally widespread resistance: concerns about the detrimental socio-ecological effects of HPPs, coupled with the substantial symbolic value attached to water, have fuelled an unprecedented – albeit fragmented – environmental movement that has united rural populations and urban environmental activists. In addition to grievances related to the adverse impacts of HPPs on ecosystems and rural livelihoods, the formal definition of (private) rights over water use and their transfer to HPP companies for 49 or 98 years by water use agreements was critical in motivating the anti-HPP mobilisations. Apart from the numerous resistance movements that erupted in dozens of localities in various regions of the country, a number of trans-local, regional and/or national networks and platforms were also formed in the struggle against HPPs, some of which are still functional and highly influential.[5]

While this does not imply that the existence and/or persistence of a struggle can be taken for granted at all HPP project sites – some were quickly suppressed, while there was no major discontent at all in a few others – the anti-HPP movement in the country warrants recognition, especially when compared to an almost non-existent counterpart in the context of dams, including mega dams that had serious environmental and social repercussions. A few cases aside, an anti-dam movement in Turkey has indeed been largely absent despite the

Turkish state's strong commitment to developing the country's hydro-potential through building dams.

The main exceptions to this have been the movement against the Ilısu Dam, which became internationally renowned, and the unrelenting resistance against the Uzunçayır Dam in Dersim. Yet these cases are highly specific examples of contestation, as their construction had been portrayed as an effort to contain particular ethnic/religious identities and their politics. The Ilısu movement was a highly diverse one that brought together transnational actors, environmentalists and the Kurdish movement, and the local opposition had primarily been rooted in framing the dam as an attack by the Turkish state to impose its power on the Kurdish population and undercut the Kurdish movement (Hommes, Boelens and Maat, 2016). The resistance in Dersim, on the other hand, mobilised the sacredness attributed to the Munzur River by the overwhelmingly Alevi population in the region and their deep-rooted (and justified) concerns about the eradication of their culture and identity (Deniz, 2016). In both cases, however, the state flatly refused these charges and instead – as has been its wont – argued that the resistance against these projects was indeed evidence of their necessity. This perverse argument was built on the premise that since hydroelectric power is essential for development and that development is a national ambition, resistance against dam building could only be explained either as a sign of ignorance and backwardness (which could be cured by socioeconomic modernisation) or as an attack on Turkey's national sovereignty (which could be beaten by extending the state's physical control over space and nature).

While we recognise that there may be a variety of reasons that underlie the double-shift in Turkey's hydro-energy landscape, the underlying link between hydropower and development has remained intact. In other words, while both the (almost exclusively) state-dam and the (almost exclusively) private sector–HPP nexus are expressions of modernisation/developmentalism construed as the national-collective interest, they imply a transformation in how nature is enrolled in the Turkish state's hegemonic project, and present different opportunities and constraints for the reproduction of state hegemony. Our aim here, therefore, is to situate the two as different configurations of state-society-environment relations within the making and contestation of state hegemony.

The imperative of modernisation and the making of state hegemony

The ambition to "catch up" with "advanced civilizations" has been the foundation of a particular political project that predates the establishment of the modern Turkish state in 1923, extending back to the eighteenth century when the decline of the Ottoman Empire in relation to European powers emerged as a political *problématique*. The diagnosis that was first made then and persisted until the current era is that this problem is a function of insufficient and/or incomplete modernisation, the cure for which is national socioeconomic development.

While ascribing pronounced priority to modernisation via economic growth is not unique to the Turkish state, its undisputed appeal is largely unmatched. Even when the very foundations of the modern republic – secularism and unitary nationalism – have been challenged by various political forces, the notion that development through rapid economic growth is a *sine qua non* for progress has remained uncontested (Adaman, Akbulut and Arsel, 2017b; Akbulut and Adaman, 2013; Arsel, 2005). Hence the country's political landscape has been dominated by debates on how best to promote economic development, where a wide range of ideologies within Turkish politics have shared the common faith in economic growth as the precondition of progress. Even when challenges to the modernisation project emerged especially after the 1980s, these critiques were not of modernisation *per se*, but rather of its top-down implementation, and at times, its strict interpretation as a replica of the Western model (Keyder, 1997).

We locate the strength and persistence of developmentalism in Turkey within the making of state and state-society relationships. In doing so, we operationalise the Gramscian notion of hegemony to highlight the amalgam of practices and discourses in which states engage to elicit active consent from society. Gramsci famously theorised hegemony as "consent backed by force" to emphasise that states do not only rule by domination, exercised from above, but seek the active support of the ruled within the sphere of civil society. Accordingly, states engage in a variety of efforts to appear impartial and justify their claim to rule, among which the constitution and reproduction of a national-popular outlook play a crucial role (Gramsci, 1971). Since the sphere of the social is always fragmented along multiple axes, eliciting the active consent of the ruled involves surpassing the diversity of interests and facilitating their unity. Constituting and reproducing a collective will (the national-popular outlook) plays a crucial role in this respect. Through such a construct, it becomes possible for the state to appear as a neutral institution that represents the general interest and justifies the dominant groups' claim to rule. On the other hand, the illusion of collective interest cements consensus among different groups in society and masks the axes of inequality that fragments the social sphere. Once an effective national-popular outlook – one that different groups in society subscribe to – is fabricated, it becomes possible for the state to appear as a neutral institution that embodies the collective interest.

We argue that, historically, the urgency to modernise and achieve economic development constituted the national-popular outlook around which a relative unity of social forces was formed in Turkey. The state achieved its power and legitimacy, first and foremost, from the promise of fulfilling the ideal of modernisation via economic growth. As Eralp (1990) has observed, this was conceived as a national endeavour, a common goal, even the questioning of which would be considered unpatriotic. The constitution of this general/collective interest also involved the portrayal of society as a homogeneous entity with no internal divisions, unified behind the goal of modernisation/development. This allowed the Turkish state to pre-empt opposition that mobilised around issues such as social justice and distribution/redistribution. There were no "classes", for instance, but rather a division of labour among the Turkish citizenry, where each and every individual worked hard to elevate the country to the level of Western civilization. The national-popular outlook served to unify different groups around a (fabricated) universal goal and prevented (albeit always temporarily) the formulation of demands arising out of intra-societal divisions. As such, the Turkish state was able to represent itself as a neutral institution that embodied the collective will and interest, and to gain the consent of its constituency.

While we hold that the imperative of modernisation/developmentalism persisted within the Turkish state's hegemonic project, we do not imply that its specific operationalisation remained unchanged historically. One major shift in this respect can be traced along the global political economic context that has been in the forefront especially since the 1980s. In a context where neoliberal policies were implemented and deepened in many parts of the world, state-society relations in Turkey were also significantly reshaped, in contrast to the previous eras marked by the state's direct involvement in the economic domain and the institutionalisation of its social-state functions. Yet we note that neoliberal restructuring rarely implies the withdrawal of active state engagement within the economic and social spheres (e.g. Madra and Adaman, 2018; Peck and Tickell, 2002), or an end of state developmentalism. Within the context of Turkey in particular, we maintain that modernisation/development,[6] albeit in different forms and guises, remains the single most important pillar on which state-society relationships depend and the strongest – yet always unfulfilled – desire that shapes the social-political imaginary.

Yet how this imaginary crystallised within the nexus of state-society-environment relations has differed across time and space. That is to say, although modernisation/developmentalism persisted as the collective interest that articulated the Turkish state's hegemonic project, its concrete manifestations within the making of environment(s) changed across particular contexts. In the rest of this chapter, we aim to trace one such change and its implications for the reproduction and contestation of state hegemony within the field of hydro-energy politics in Turkey.

Hegemony reproduced and contested: dams versus HPPs

In the context of hydro-energy, the modernisation/development imperative translated into developing the country's hydro-potential to the utmost level, where the rhetoric of "idle" or "raging" rivers flowing away – instead of being exploited to fuel growth through irrigation or power generation – was continuously invoked. Juxtaposed to this has been the image of the state as the omnipotent deliverer of modernisation and economic development, which tames nature and/or reveals its economic potential for the benefit of the whole (unified) nation. "The king of dams" Süleyman Demirel is perhaps the epitome of this strategy, who served as the prime minister multiple times between 1965 and 1980 and after his post as the head of the State Hydraulic Works (SHW). Demirel completed a couple of dozens of dams and initiated the construction of more than 50 during the period he held office, in addition to launching the

102   *Bengi Akbulut, Fikret Adaman and Murat Arsel*

gigantic SAP. His words regarding the SAP are illustrative of the significance attributed to dams within state-society relationships:

> the love of the SAP is the love of Turkey. The SAP is the cement that unifies Turkey; it is the largest project of the Republic. It is one the biggest projects being undertaken today in the world. It is beyond an engineering project. In our terminology, that is to say the terminology of scientists, of technicians, it is an integrated project. And it has the human at its centre. What does this mean? This means that it is about rescuing the human from being trapped between the cracked soil and the blue sky. It is about easing humans' struggle to live.
>
> It is a struggle to make people happy. It is not only about taking water from rivers and bringing them to plains. That is just a part of the big picture. It includes the education of people, their preparation for a new world, for the conditions of a new world. It includes fertility, fullness [of the stomach] and prosperity.[7]

Demirel's words attest to a vision of progress and national unity that went beyond merely *economic* development and pointed to a greater transformation in Turkish society. They exemplify a vision of dams through which the state promises and/or delivers a "new world" of modernisation to society, and thus point to the role that dam building has played in the making of the state in Turkey. Indeed, dams and associated large hydropower schemes have unexceptionally been cast as key contributors to accelerating Turkey's quest for modernisation, which surpasses a solely economic transformation process despite the central role given to economic growth within it (Adaman and Arsel, 2010). This was buttressed by the constant highlighting of their supposed contribution to local employment and economic development, in order to brush aside their destructive social and ecological impacts.

We argue that constructing dams has been a constitutive practice of the state's hegemonic project in Turkey. That is to say, dams emanate the image of the state as the embodiment of society's collective interest, through which the active consent for its claim to rule can be achieved. To the extent that modernisation/development was construed as the collective interest, the contribution of dams to modernisation, economic development or employment was represented as benefits to be equally shared among the citizenry. The promise of modernisation through dams was thus worked to create the image of a homogenous social sphere and veil the socioeconomic fragmentations within it, and reinforced the representation of the state as an *impartial* and *legitimate* actor embodying the collective interest. The monumental appearance of the dams (some of which are featured among the highest dams in the world) and the technological requirements to construct them further reproduced the existence of the state in a most visible way and materialised the very ideal of modernisation/development. In that sense, they are powerful symbols that both concretise and reinforce the hegemonic project of the state (Menga, 2015; Mitchell, 2002).

While the detrimental social and ecological impacts of dams (to the extent they were acknowledged) were portrayed as unavoidable burdens on the path to development, the injustices they created – including forced resettlements in addition to socio-ecological destruction – were depicted as inevitable compromises that all citizens had to make for "the greater good". This greater good was of course development, which was seen as necessary not only to fulfil the "yearning" of Turkish citizens for such projects but also to advance the cause of aligning Turkey with other "advanced civilizations". Demirel, once again, made this point at the unveiling of the Keban Dam in 1965 when he spoke as follows:

> [E]very stage of the struggle for civilization unfolds the same way: It involves ardour and sweat; it is gruelling, and it comes with hardship. . . . Like [all such endeavours], in addition to the great possibilities the Keban Dam will make available, there will also be some inconveniences.

Explaining that as the reservoir of the dam fills up some villages in the area will be inundated and forced to relocate, Demirel acknowledged that the residents of these villages were uneasy at the prospect of losing their homes and fields. "There is no need for them to be uneasy," he added. "As the state creates a civilizational work for them, it will not disadvantage them under any circumstances."[8]

This implied distribution of mutual sacrifices across society resonates closely with the conception of a division of labour among the unified citizenry to achieve the collective interest of modernisation/development (see above). In this sense, dams became a tool not only to concoct general interest in society that transcended the internal fragmentations within it, but were also operationalised to cement the illusion that there were no divisions within society and that every citizen was doing their part for development.

On the other hand, the *inevitability* and *mutuality* of compromise for the sake of modernisation/development was coupled with an implicit or explicit notion of *reciprocity*, not only by the presumed diffusion of the benefits of modernisation/development, but also by the various mechanisms of compensation offered by the state. That irrigation services will be extended or other forms of public infrastructure will be provided in order to compensate for the destructive effects of dams are just a couple of examples. Mobilising reciprocity as such does not only reproduce the image of the (benign) state working hard for its people (service provision, infrastructure, irrigation), but also – more importantly – situates dams within a specific notion of fairness. That is to say, (promising) the provision of public infrastructural services is not simply an act of economic compensation (or "bribing" locals into agreement) based on some form of calculative behaviour of the agents involved. Rather, it gains significance as the Turkish state's claimed impartiality in mediating grievances and distributing the benefits of development within the specific constellation of the state-society relationship: while every entity in the country is to make compromises for the sake of general interest operationalised by the state, all will receive the fruits of their sacrifices, not only through the realised objective of modernisation/development, but also in the form of regional development, public services, employment, etc. distributed by the state. The state is thus positioned as an impartial actor that mediates who will make which sacrifice and who will receive which fruits within the operationalisation of developmentalism/modernisation as a national-popular outlook.

Our claim is neither that the Turkish state has in fact been a fair or an impartial mediator of the developmental process, nor that the notions of reciprocity and mutuality have been genuine. Rather, we argue that dams, together with the practices and discourses produced around them, were positioned within a particular state-society relationship that animates these notions as the pillars of the state's hegemonic project. In that sense, it might matter little whether compromises towards the collective interest were actually reciprocated or mutually shared across society, as these notions gain their strength primarily from the fact that they are *promised*, *imagined* and/or *expected* to be fulfilled. In turn, such positioning of dams as a form of hydro-technology in general has effectively mobilised local consent for dam projects, despite their social and ecological costs and the particular distribution of such costs. We claim that it is precisely such a positioning that is missing in the case of HPPs.

This does not imply that mechanisms to elicit local consent for HPPs were absent. The widespread and vocal opposition to HPPs erupted despite efforts by energy companies to implement a variety of local compensation schemes as consent-building mechanisms. Typically, energy companies build local infrastructure such as schools and hospitals, improve roads and provide social assistance in the form of scholarships for local children or various in-kind contributions to local communities. HPPs were also strongly endorsed by the Turkish state as indispensable components of modernisation/development, and were represented not only as the keystones of this imperative but also as the very indicators of its realisation. The following statement by Erdoğan, the then prime minister, became (in)famous as an exemplar:

> We are obliged to make efforts to produce energy by our own resources, either through dams or HPPs. We are still trying to handle this by natural gas cycle plants. I should remind my citizens how costly this is. Both in households and in industry, it lowers our competitiveness and threatens us. That means we need to acquire energy more cheaply so that our competitive power can increase. This is because Turkey is now a leaping, roaring country in industrialisation, in technology. That means we need to facilitate the steps our country is taking by these infrastructural works and strengthen our competitive standing with the world.
>
> A country's consumption of electricity also signifies the level of development, the productive capacity of that country. We cannot overlook that. The more a country consumes electricity, the stronger it is, the faster it advances on the path to development. It means that the gears in the factories are turning, that production in our enterprises is rising, that its consumption in the households is increasing, that technology use is spreading across the entire country.
>
> . . . This is why we are taking a new step; we are replacing the phrase "water flows, the Turk just watches" with "water flows, the Turk acts", and God willing, we are compensating for this shortage.[9]

Yet the dynamics of consent that rendered dams a constitutive practice of society's collective interest remained absent in the case of HPPs. While the appeal and legitimacy of modernisation/developmentalism as a collective interest might (and indeed does) go unchallenged, its specific operationalisation through different hydro-technologies might be (and indeed is) contested. We argue that as a form of hydro-technology, HPPs failed to mobilise reciprocity and mutuality in particular, through which the ideal of modernisation/developmentalism could be articulated as a collective interest. In that sense, we point to the multi-layered and multi-actored nature of hegemony at the local scale (see also Loftus and Lumsden, 2008). The making of local environments through different forms of hydro-technologies is situated and thus substantiated within the broader constellation of state-society relationships. Yet the extent to which local processes of consent and hegemony can be mobilised through these interventions depends on how effectively they can (re)articulate the notion of a collective interest.

The next section illustrates the arguments made so far through a case study we have conducted in Artvin, located in the Çoruh Basin in the north-eastern corner of Anatolia. The provincial region of Artvin hosts dozens of HPPs that are operational, under construction or in the planning stage; there are also three large operational dams and several more will become operational in the near future. Our study focuses on the Ardanuç-Şavşat-Borçka triangle in the north of the province, where we conducted the fieldwork intermittently between 2012 and 2014.

The Çoruh Basin: shifting hydropower landscapes

The Çoruh River Basin, with a total catchment area of 22,000 square kilometres extending beyond the Georgia border, holds not only high ecological and historical value but also immense hydro-potential, which drew the Turkish state's developmentalist gaze long ago. Preliminary studies to develop the hydro-energy potential of the Çoruh Basin were initiated in the early 1960s, culminating in the Çoruh River Basin Development Plan that was made public in 1982. The plan foresaw the construction of 10 dams on the river's main body and five on its tributaries, in addition to 17 run-of-river plants, over the course of two decades (see Figure 7.1). Of these, the three dams that are operational today, Muratlı, Borçka and Deriner, were completed in 2005, 2007 and 2013, respectively. Construction of the Yusufeli Dam, which received wide public attention due to its large resettlement impact (submerging Yusufeli town centre as well as 19 surrounding villages and displacing 20,000 people), only began in 2012 and is expected to finish by 2019.

The elongated time frame of the Plan coincided with various phases of energy market liberalisation (Evren, 2014). Consequently, while construction of the first three dams (Muratlı, Borçka, Deriner) was undertaken solely by the SHW, foreign investment was sought through bilateral agreements for the Yusufeli and the Artin dams (mostly due to their colossal sizes) and the remaining five projects in the upper basin were transferred to private sector investments in the early 2000s (Eroğlu, 2013; Evren, 2014).

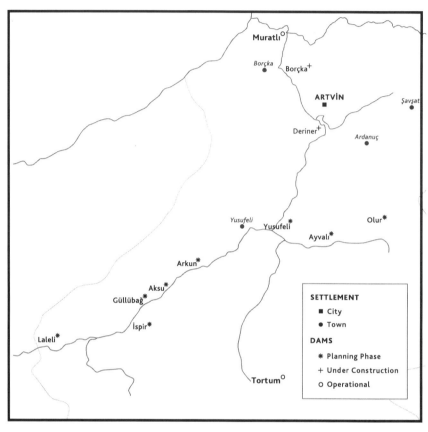

Figure 7.1 Çoruh Basin Project
(Source: www2.dsi.gov.tryusufeli_projesi.pdf)

Weighed against the Project's ambitious size and scope, the local contestation it triggered remained meagre at best. The local movement initiated in Yusufeli was the only significant one, even though it died off in the late 2000s after about 10 years of rallies, protests, lawsuits, networking and lobbying against European credit agencies (Evren, 2014). Evren (2014), tracing the rise and fall of the anti-dam campaign in Yusufeli, argues that the movement lost its original momentum after the dam's re-nationalisation in 2006, where state endorsement of the project (rather than an international consortium) translated into a certain expectation (and faith) that not only (development) damages would be compensated, but desires and aspirations of locals more broadly *vis-à-vis* development would finally be fulfilled. The short-lived opposition in Borçka, on the other hand, was quickly co-opted via promises of local employment in the dam construction (Evren, 2014). These two cases aside, the sole source of conflict around the Çoruh Project seems to have been about the amount of settlement/compensation paid by the state, and the anti-dam mobilisation in the region was feeble (Aksu, 2017; Evren, 2014).

Against the little contestation that the dam projects have sparked, the last decade's widespread and highly visible resistances spurred by HPPs pose a contrast. The latest round of Turkey's energy liberalisation brought 105 planned HPPs to Artvin, of which 19 are operational and eight are under construction as of today, in addition to more than 20 that are in various stages of licensing[10] (see also Aksu, 2017). While the prevalence and strength of the mobilisations against this type of hydropower investments are by no means homogenous across the region, their visible and vocal existence is undeniable. In particular, within the Ardanuç-Şavşat-Borçka triangle that comprises our present focus, a dynamic opposition has existed since the late 2000s. In addition to constant organising through numerous meetings and rallies by local environmental groups, locals often took direct action – e.g. driving company employees and accompanying security forces away in their struggles (Hamsici, 2010). The majority of the HPP projects have been challenged via legal means as well; some court cases resulted in victory for the local movements, while dozens of others are still ongoing.

Although some of the reasons that motivate the anti-HPP resistance movements in the region relate to aspects specific to HPPs as a particular form of hydro-technology – reduction of dissolved oxygen concentrations in water, constraints on small-scale irrigation, etc. – it is hard to claim that they are environmentally, economically or socially more destructive than dams. That is to say, the emergence of a robust and strong environmental resistance movement against HPPs seems to have less to do with their experienced and expected impacts, and more with how these impacts are perceived, interpreted and positioned, in particular within the constellation of state-society relationships.

Echoing the grievances raised more broadly by the anti-HPP resistance movements across the country, the main underlying reason for opposing HPPs in the region has to do with their (actual or expected) impacts on livelihoods. Reduced water availability hurts irrigated farming, and becomes a serious concern especially for villages with arable land and direct marketing links with town centres;

heightened noise and pollution caused by construction and related developments have adverse effects on beekeeping, a prevalent source of livelihood especially in Borçka and Ardanuç. Oppositional motivations go beyond livelihood concerns to encompass a defence of "living spaces" – a particular type of social relations structured around a broad understanding of the rural commons and a robust stance against the privatisation of water.

On the other hand, certain aspects specific to HPPs, such as the introduction of water use rights as a form of (temporary) private property right to water and appropriation of streams at their source, have also mobilised considerable dissent that fuelled anti-HPP mobilizations. However, such grievances were typically situated within perceptions of private/particular interests that were attached to HPPs (as opposed to the collective interest undergirding dams), and became motivations for resistance through the lens of these perceptions, as will be discussed below.

Layers of local hegemony: consent and contestation in Çoruh Basin

> Consider, for instance, the slogan "electrification is socialism". . . . Anti-imperialism and socialism are still interpreted as some form of developmentalism and statism. The most crucial impact of this [mode of thinking] is the lack of reaction to the dam issue in Artvin. There are many educated people in Artvin, literate in politics. The worst you'll find are CHP [Republican People's Party][11] supporters. Artvin, Şavşat, Ardanuç have all bred many revolutionaries . . . But no one made any fuss about the dams. Why? [Because] [d]ams mean electricity, electricity means industrialisation, and industrialisation means development. It's as simple as that! Maybe there are some side issues as well but basically this is the gist of it. Neşe [a prominent figure within the local ecological movement] says the same thing in the interview we did for our journal. She says that they initially approached the [dam] issue as a matter of development.
>
> (Aksu, 2013: 43)

> Since mines are dug on top of the soil, illustrating and explaining the destruction had been relatively easy. But people were clueless about dams until then, so we could not explain [their potential impacts]. And the state managed to sugar-coat it really well. We were aware of the importance of this valley, but even we could not grasp that a man-made structure [such as a dam] can be so destructive. Even those on the Board [of Green Artvin Association – *Yeşil Artvin Derneği*] were convinced that we should not oppose the dams, that the state needs energy.
>
> (Karahan, 2013: 49)

These quotes, by two prominent figures from the local environmental movement in the Artvin region, demonstrate how the general notion of a dam effectively operationalises the ideal of modernisation/development across different groups from a variety of ideological backgrounds, including environmentalists. They attest, in particular, that it is precisely the close resonation of dams with this imperative, constructed as the collective interest, which mobilised local consent for the construction of the large dams in the Çoruh Basin.

While these quotes speak to the strength of developmentalism as an unquestioned imperative and a societal goal, they do not imply that all projects embedded within the imperative of modernisation/development mobilise local consent automatically. The extent to which this imperative can be constituted as a collective interest within a specific project becomes critical in terms of eliciting consent, and differs across dams and HPPs in particular. The contrast between dams and HPPs in this respect was echoed in a statement made repeatedly by locals: "Dams are different, they are [built] for the whole country's electricity". That a collective/general interest was identified with dams is more clearly delineated when juxtaposed with private interests associated with HPPs, as they were in the following statements:

> When the state said "I am going to put a golden bracelet on Çoruh", we gave [our water], we did not say anything. We did not oppose the dams. But dams are not built so that someone profits, so that someone takes and sells our water. [HPPs] are a different thing entirely. They will take our water from here and produce electricity, and then sell it to who knows whom.
>
> (villager, Ardanuç)

> If the issue is to produce energy, we have already given [up] our water. Dams have been built on the main tributaries of Çoruh. Now they want to put clamps on its capillaries. We should not be asked to give up anything else. But the [real] issue is not [about] energy. Water resources will be depleted in 2020. And then the companies who have acquired the water use rights for HPPs will sell the water.
>
> (villager, Borçka)

Identifying dams with the collective interest and HPPs with private motives was also implied in how the socio-ecological impacts of the projects were understood and evaluated. Generally speaking, the impacts of dams were considered more tolerable, while those of HPPs – even if less critical – were highly disputed, precisely because they were seen as being for "someone else's profit". This was paralleled in the context of land expropriations associated with the projects. Although land expropriation is sanctioned by state law and carried out by state organs in both cases, expropriations for HPPs were contested on grounds that they involved the *private* interest of the companies while expropriations for dams were overwhelmingly perceived as necessary acts and thus legitimate. Furthermore, the dams built in the Basin were often implicitly or explicitly associated with mutuality, which reinforced their perceived legitimacy. The local burden of the dams was put on par with sacrifices made both by communities in different parts of Turkey, and by different generations for the development and progress of the country.

The locals often linked the disparity between the idea of a collective interest mobilised by the dams and the perception of private interests associated with HPPs to a particular notion of fairness embodied by the former and missing in the latter. For instance, the adverse impacts of dams were stated to be "more defined and bounded", because they were kept to a minimum, as the state was trying to "strike a balance between developing the country and not aggrieving people [too much]" (villager, Şavşat). HPPs, on the other hand, were perceived to be more destructive (many consecutive plants can be built on a single stream and effectively appropriate the entire body of water) since the HPP companies were acting by the sole motivation of (private) profit-making and had no incentive to observe fairness.

In a parallel vein, dams were often seen to have operationalised the notion of reciprocity more effectively. The compensation paid for the dams (for land expropriations and resettlements) were assessed as fair, and employment benefits were found more satisfactory in general, while HPPs were typically described as "giving nothing back". The expectation of reciprocity for local consent is apparent in the words of a villager from Borçka, "If HPPs will be built than we should be provided with electricity, with water. Or we should be given a tax break". Yet the notion of reciprocity evoked here goes beyond a mere compensation mechanism – such as the social responsibility projects that were implemented by the HPP companies – and appeals to an equitable sharing of the benefits of modernisation/developmentalism. The following statement by a local mayor puts it even more succinctly:

> When [the HPP company representatives] came here, I told them I was against it. I told them that they should not be built in my opinion; they would destroy the whole area. [He mentions the example of Papart Valley in Şavşat.] But if there is no escape from them, I told them that at least 10–15% of the revenue should be returned to the villages. . . . They will take everything from the villages, they will take the water, and not give anything back. This is unacceptable.

Finally, while the overarching imperative of modernisation/development was not challenged even within the context of HPPs, the shift in state-society relationships around the changing form of hydro-energy in the region was being questioned, albeit modestly. The following statement by the same mayor encapsulates the limits to which the Turkish state can maintain and reproduce its claim to embody the collective interest through HPPs:

> They had to reduce the conservation status area in order to build the HPP on the road exiting from [the town centre]. They are obligated [by law] to obtain an evaluation report from the municipality in order to do that. That is how we were informed. [If there was no legal obligation] we would never know. They will come here and build an HPP, yet they can't be bothered to inform anyone, neither the local people nor the municipality. There is no such obligation if it was not for the conservation area. This is not an administration that governs for its citizens, that makes something together with them, but rather an administration that is hostile to its citizens.

To recapitulate, while the broader imperative of modernisation/developmentalism goes largely unchallenged in the Çoruh Basin, the specific forms of hydro-technology through which this imperative was operationalised carry different implications in terms of producing local consent. That is to say, HPPs were not opposed primarily or necessarily because the ideal of modernisation/development does not function as a collective interest, but rather because they failed to animate this ideal effectively despite the Turkish state's reiteration of their role to fulfil the collective interest of modernisation/developmentalism. Within this context, the extent to which different forms of hydro-technology mobilise understandings of fairness, mutuality and reciprocity emerge as a critical aspect of consent. While the adverse socio-ecological impacts of dams were interpreted within such understandings and found legitimate, the absence of such a positioning in the case of HPPs fail to locate them as constituents of the collective interest.

Conclusion

This chapter problematised the relative absence of social contestation in the face of large-scale dams when juxtaposed with the vocal and widespread opposition to HPPs in Turkey. In doing so, it located different forms of hydropower interventions within the broader constellation of state-society relationships through a Gramscian framework. The chapter, in particular, demonstrated the varying extents to which different forms of hydropower can operationalise modernisation/developmentalism as a collective interest that elicits consent from a broad section of society.

The ability of the idea of development – one that is expressed in terms of a linear process of increased accumulation – to create consent remains largely undiminished in much of Turkey (with the possible exception of certain segments of the Kurdish community). However, development – and the manner in which its promise and potential was deployed by the state – failed to generate the necessary societal consent in the case of HPPs – and not because their environmental impacts are more dramatic than those of large-scale dams. This chapter argued that what sets HPPs apart from large-scale dams in terms of societal conflict is the manner in which they have been put into operation, rather than any differences in developmental promise or environmental impact.

More specifically, HPPs failed to embody the appearance of collective interest at whose altar affected communities would abandon their grievances. Their failure to appeal to collective interest, in turn, was a function of the neoliberal political economic context in which HPPs were unleashed on the countryside. While they too were supposed to help Turkey move forward in its modernisation process, the manner in which HPPs were planned and constructed, as well as the politico-legal processes through which they were rendered legitimate, undermined the notions of reciprocity and mutuality, which had helped portray large-scale dams as necessary or even desirable for the sake of the nation's collective interest, despite the heavy local environmental and social tolls they exacted.

112 *Bengi Akbulut, Fikret Adaman and Murat Arsel*

It is important to highlight the epistemological challenge that is central to this chapter. Whereas much of contemporary critical development studies and political ecology research have focused on – it could even be argued, fetishised – various forms of open, active and easily observable conflicts, our goal here was to contrast the absence of societal resistance against large-scale dam projects with the emergence of vigorous contemporary struggles between state, society and private capital against HPPs. Doing so is not only methodologically worthwhile, for our understanding of conflict dynamics would arguably be partial if it were only built on cases where a particular type of conflict is present. It is also important as it would help separate proximate causes from underlying structural dynamics in cases where open conflicts are present.

Notes

1 We would like to thank Begüm Özden Fırat and the editors of this volume for their comments on earlier versions of this chapter. Umut Kocagöz provided much appreciated research assistance. The usual caveats apply.

2 The first round of the field study was conducted in October 2012, followed by two more rounds in August–September 2013 and May 2014, respectively. The field study comprised of in-depth interviews held with local villagers and environmental activists. Villages that are against HPPs as well as those that support HPPs were visited. A total of 27 interviews were held. We would like to thank Umut Kocagöz for providing valuable research support during the field study.

3 Data obtained from the Southeast Anatolian Project official website: http://gap.gov.tr [accessed 10 May 2017].

4 This is not to suggest, however, that the state has withdrawn from energy generation altogether. Rather, it has been shifting its target to other energy sources, especially nuclear power (see Akbulut, Adaman and Arsel, 2017).

5 Water Assembly (*Su Meclisi*), The Fellowship of Rivers (*Derelerin Kardeşliği*), Black Sea in Rebellion (*Karadeniz İsyandadır*), Platform Against the Commercialisation of Water (*Suyun Ticarileşmesine Hayır*) and Green Rage (*Yeşil Öfke*) are among the prominent networks and platforms formed against HPPs.

6 We use modernisation/development interchangeably throughout the text to highlight the fact that the two have come to mean the same thing, both in the eyes of the state and in the social imaginary.

7 *Demirel' in Sevdası: GAP (Demirel's Passion: SAP).* http://arsiv.sabah.com.tr/1997/05/30/r11.html [accessed 11 May 2017]; translation ours.

8 *Başbakan Süleyman Demirel'in Keban Barajı Temel Atma Töreni Konuşması (Prime Minister Süleyman Demirel's Speech at the Groundbreaking Ceremony of Keban Dam).* www.imo.org.tr/resimler/ekutuphane/pdf/8096.pdf [accessed 10 May 2017]; our translation.

9 *Başbakan Erdoğan, İkizdere'de HES'i Hizmete Açarken Çevrecilere Çattı (Prime Minister Erdoğan Takes a Swipe at Environmentalists at İkizdere HPP's Opening Ceremony).* www.milliyet.com.tr/basbakan-erdogan-ikizdere-de-hes-i-hizmete-acarken-cevrecilere-catti-siyaset-1275291/ [accessed 10 May 2017]; our translation.

10 Of these, 25 are in Yusufeli, 21 in Borçka, 21 in Şavşat, eight in Murgul, 14 in Arhavi, two in Hopa, nine in Artvin's provincial centre and five in Ardanuç. Data obtained from the SHW: http://bolge26.dsi.gov.tr [accessed 10 May 2017].

11 Republican People's Party (*Cumhuriyet Halk Partisi*), currently the main opposition party in parliament, represents the majority of social democrats in Turkey. It also became the main political outlet for former radical leftists after the military coup of 1980, especially in the peripheral regions of Turkey.

References

Adaman, F., Akbulut, B. and Arsel, M. eds. 2017a. *Neoliberal Turkey and its discontents: Economic policy and the environment under Erdoğan.* London: I.B. Tauris.

Adaman, F., Akbulut, B. and Arsel, M. 2017b. Introduction. In Adaman, F., Akbulut, B. and Arsel, M. eds. *Neoliberal Turkey and its discontents: Economic policy and the environment under Erdoğan.* London: I.B. Tauris, pp. 1–17.

Adaman, F. and Arsel, M. 2010. Globalization, development, and environmental policies in Turkey. In Çetin, T. and Yılmaz, F. eds. *Understanding the process of institutional change in Turkey: A political economy approach.* New York: Nova, pp. 319–335.

Akbulut, B. and Adaman, F. 2013. The unbearable charm of modernization: Growth fetishism and the making of state in Turkey. *Perspectives: Political Analysis and Commentary from Turkey.* **5**(13), pp. 1–10.

Akbulut, B., Adaman, F. and Arsel, M. 2017. The radioactive inertia: Deciphering Turkey's antinuclear movement. In Adaman, F., Akbulut, B. and Arsel, M. eds. *Neoliberal Turkey and its discontents: Economic policy and the environment under Erdoğan.* London: I.B. Tauris, pp. 175–190.

Aksu, C. 2013. Röportaj: Doğa ile doğalcılık arasında Karadeniz ekoloji mücadelesinin açmazları [Interview: Shortcomings of the Blacksea ecological struggle in between the nature and naturalism]. *Kolektif Dergi.* **16**, pp. 41–47.

Aksu, C. 2017. *Artvin'de HES'ler: her şeye rağmen 3–5 ağaç meselesi [HPPs at Artvin: Despite everything, struggle for 3–5 trees].* www.artvinonline.com/2017/04/artvinde-hesler-her-seye-ragmen-3-5-agac-meselesi/ [Accessed 10 May 2017].

Arsel, M. 2005. Reflexive developmentalism? Toward an environmental critique of modernization. In Adaman, F. and Arsel, M. eds. *Environmentalism in Turkey: Between democracy and development?* Aldershot: Ashgate, pp. 15–34.

Çarkoğlu, A. and Eder, M. 2003. Domestic concerns and the water conflict over the Euphrates-Tigris river basin. *Middle Eastern Studies.* **37**(1), pp. 41–71.

Çarkoğlu, A. and Eder, M. 2005. Development *alla Turca*: The Southeastern Anatolia project. In Adaman, F. and Arsel, M. eds. *Environmentalism in Turkey: Between democracy and development?* Aldershot: Ashgate, pp. 167–183.

Deniz, D. 2016. Dersim'de su kutsiyeti, Mizur/Munzur nehri ilişkisi, anlamı ve kapsamı ile baraj/HES projeleri [Hydro-energy, water, and faith along the Mizur/Munzur River, Dersim]. In Aksu, C., Erensü, S. and Evren, E. eds. *Sudan sebepler: Türkiye'de neoliberal su-enerji politikaları ve direnişleri [Watery reasons: Neoliberal water-energy politics in Turkey and resistance].* Istanbul: İletişim, pp. 177–197.

DSI (Devlet Su İşleri) 2016. *Faaliyet raporu [Annual report].* Ankara: Devlet Su İşleri Genel Müdürlüğü.

Eralp, A. 1990. The politics of Turkish development strategies. In Finkel, A. and Sirman, N. eds. *Turkish state, Turkish society.* London: Routledge, pp. 219–258.

Erensü, S., Evren, E. and Aksu, C. 2016. Giriş: Yeğin sular daim engine akar [Introduction: Strong streams always go to the horizon]. In Aksu, C., Erensü, S. and Evren, E. eds. *Sudan sebepler: Türkiye'de neoliberal su-enerji politikaları ve direnişleri [Watery reasons: Neoliberal water-energy politics in Turkey and resistance].* Istanbul: İletişim, pp. 9–33.

Eroğlu, V. 2013. *Çoruh'un mavi gerdanlıkları [Çoruh's blue necklace].* Ankara: Devlet Su İşleri Vakfı.

Evren, E. 2014. The rise and decline of an anti-dam campaign: Yusufeli Dam Project and the temporal politics of development. *Water History.* **6**(4), pp. 405–419.

Gramsci, A. 1971. *Selections from the prison notebooks of Antonio Gramsci.* New York: International Publishers.

Hamsici, M. 2010. *Dereler ve isyanlar* [*Rivers and uprisings*]. Istanbul: Nota Bene Yayınları.

Harris, L. 2002. Water and conflict geographies of the Southeastern Anatolia Project. *Society and Natural Resources*. **15**(8), pp. 743–759.

Hommes, L., Boelens, R. and Maat, H. 2016. Contested hydrosocial territories and disputed water governance: Struggles and competing claims over the Ilısu Dam development in southeastern Turkey. *Geoforum*. **71**(May), pp. 9–20.

Karahan, N.N. 2013. Röportaj: Doğa gidince altınları mı yiyeceğiz? [Interview: Are we going to eat gold when the nature comes to an end?]. *Kolektif Dergi*. **16**, pp. 48–51.

Keyder, Ç. 1997. Whither the project of modernity? Turkey in the 1990s. In Bozdoğan, S. and Kasaba, R. eds. *Rethinking the project of modernity in Turkey*. Seattle: University of Washington Press, pp. 37–51.

Loftus, A. and Lumsden, F. 2008. Reworking hegemony in the urban waterscape. *Transactions of the Institute of British Geographers*. **33**(1), pp. 109–126.

Madra, Y.M. and Adaman, F. 2018. Neoliberal turn in the discipline of economics: Depoliticization through economization. In Cahill, D., Cooper, M., Konings, M. and Primrose, D. eds. *SAGE handbook of neoliberalism*. California: Sage, pp. 113–128.

Menga, F. 2015. Building a nation through a dam: The case of Rogun in Tajikistan. *Nationalities Papers*. **43**(3), pp. 479–494.

Mitchell, T. 2002. *Rule of experts: Egypt, techno-politics, modernity*. California: University of California Press.

Özok-Gündoğan, N. 2005. "Social development" as a governmental strategy in the Southeastern Anatolia Project. *New Perspectives on Turkey*. **32**(Spring), pp. 93–111.

Peck, J. and Tickell, A. 2002. Neoliberalizing space. *Antipode*. **34**(3), pp. 380–404.

8 An island of dams

Ethnic conflict and the contradictions of statehood in Cyprus

Panayiota Pyla and Petros Phokaides

A typical view from tourist apartments in coastal cities of Cyprus shows both pools on the ground and water tanks on the roofs, capturing a powerful contradiction of contemporary Cyprus. On the one hand, water consumption is extravagant, with private pools in residential areas, not to mention the golf courses and resorts all around the island, as locals emulate the lifestyle advanced by the vigorously growing hotel industry. On the other hand, water shortage is so severe, and water supply has been so erratic, that water tanks have become a fixture on the roof of dwellings, a quintessential emblem of this drought-stricken Mediterranean region.[1] Since the 1960s, the government has been attempting to tackle the island's drought problem with big dams, which have earned Cyprus the uncomfortable distinction of being 'the most dam-dense country in Europe'; by the 1990s, the government's strategy shifted from dams to desalination plants. Despite these efforts, the island is ranked among the top twenty water-scarce countries of the world (Kotsila, 2010: 10). Domestic water supply is still a major problem, and people continue to feel the need to store water on their roofs.

Cyprus is probably less known for its dams than for its intercommunal conflict and its ethnic division. This so-called "Cyprus problem," already identified as 'insoluble' (Woodhouse, 1958: 193), was aggravated in 1963 when the Turkish Cypriots pulled out of the country's bi-communal government, three years after the island's independence from the British. The Turkish Cypriots withdrew into enclaves, leaving the Greek Cypriots, or 80% of the total population, in total control of governmental operations. Intercommunal talks began in 1967, but the "Cyprus problem" remains to this day, even if it has since been transformed in various ways, earning the island another dubious distinction: 'a world nuisance,' according to one British humourist (cited in Ker-Lindsay, 2011: xii). All these developments happened against the backdrop of Cold War geopolitics, with Britain, Greece and Turkey keeping a particularly close eye on the island's internal politics.

Considering the political and spatial realities sketched above, this chapter examines the relationship of water – its access and management – to the long intercommunal conflict on the island. The main question driving this study is: How have strategies for combating recurrent drought become entangled with ethnic tensions, the creation of enclaves, and even foreign military intervention?

And how did the island's ambivalent geopolitical allegiances play a role in water development decisions? To what extent has the profuse building of dams and other water infrastructures supported (or not supported) aspirations to peace-building?

This chapter tackles these questions by investigating the recent history of water management on the island of Cyprus, and by assessing, for the first time, the complex relationships between water infrastructure development and the processes of nation-building and peace-building in Cyprus, against the background of a larger geopolitical context. It focuses on the programme of intensive water management in the 1960s, a programme funded by the United Nations (UN) and wholeheartedly supported by the Republic of Cyprus, and demonstrates that the political management of the "Cyprus problem" that began about the same time was intricately connected with water management in ways that current discussions on water do not recognise. The case of Cyprus, we argue, uncovers nuances in the relationship between the techno-scientific and the political, and exposes unique connections between development and peace-building rhetoric.

Post-independence Cyprus and its water-dependent future

In 1960, the year the Republic of Cyprus was officially inaugurated, the UN sent a seven-member mission to Cyprus to evaluate the economic potential of the new island nation and to make recommendations for economic development. The outcome of this assessment was the so-called "Thorp report," named after the chair of the mission who authored it (Thorp, 1961). Willard Long Thorp (1899–1992) was a professor of economics who had served the United States as a government official in various posts during the implementation of the Marshall Plan aid to Western Europe. In the early years of the UN, he had also been in charge of 'technical assistance to industrially underdeveloped countries' ('Willard L. Thorp', n.d.). As a UN expert closely affiliated with the United States government, Thorp was committed to the political and ideological agendas of the United States' assistance programme in the post-war era. In his extensive writings from the early 1950s, he argues that the aid to 'underdeveloped' countries must 'create situations of political and economic strength,' and that with such help they would create 'resistance . . . to the coming to power of Communists or seriously hostile governments.' He asserted, in fact, that, 'in the present world contest, every country is important' (Thorp, 1951: 416), revealing the tenacity of commitment and the comprehensive breadth the United States and the UN required in their assistance policies.

Thorp's report for Cyprus highlighted the need to 'accelerate the process of economic growth' (Thorp, 1961: 1), superimposing onto the island his larger geopolitical understanding of "help" to "underdeveloped" countries as a strategy to enable economic and political stability against Soviet influence. Indeed, despite the country's small size, Cyprus and its development had been widely recognised in United States policy circles as an 'important segment of the global antagonism between capitalism and communism' (Nicolet, 2002: 96). Echoing such perceptions, the Thorp report strove to create a comprehensive basis for long-term development in Cyprus, forming an authoritative reference for several of the government's later five-year plans. The UN also set its expectations very high, charging the Thorp report with the task of guiding not only the local government, but also, according to an official press release, 'all countries and Agencies interested in assisting Cyprus with development programmes' (Press and Information Office (PIO), 1962).

According to Thorp's ambitious vision, the key to Cyprus's long-term and comprehensive development was 'the effective marshalling of resources and their efficient use' (Thorp, 1961: 2). The report immediately cast a spotlight on water, which was perceived as 'the key natural resource'; the productivity of 'land, labour and capital investment' was tied to adequate water resources (Thorp, 1961: 6).[2] Following a development expert's typical (if paradoxical) view of the time, Thorp called for more resource surveys, even as he had already decided that 'natural resources . . . can be increased' (1961: 1). This contradictory technocratic drive for increasing resources may have been taken for granted during the heroic era of international development, but why, in Thorp's mind, should Cyprus increase its water resources? How did he think it should increase them? And who could increase these resources? The answers to these questions contribute to the assessment of both the impact of international development expertise in the 1960s and its current repercussions on Cyprus.

The increase in water resources was deemed necessary because agriculture was singled out as 'the most important economic field' (PIO, 1962), and water-intensive agricultural production in particular, because it yielded high-price agricultural goods for export. From Thorp's developmentalist point of view, water-intensive agriculture was a clear priority because it gave maximum economic returns and expedited economic growth. The fact that water-intensive agriculture would also intensify water dependence in a water-scarce locale was circumvented as an issue that could be managed, presumably, through water development.

This is why Thorp's report went so far as to assert that 'with appropriate development, water should no longer be a limiting factor on agricultural production' (1961: 7). In hindsight, the optimism that water could eventually be so abundant that access to it would not be a limiting factor in agricultural output may seem stunning; however, the reality of finite resources at the time was bypassed in favour of foreign expertise. What is most striking is not the hubris of the development expert, who in fact was proved wrong, but that by the 1990s, when water storage capacity was indeed maximised, the acute droughts made it clear that the dam programme would not solve Cyprus's water problem.[3] The issue is *also* that Thorp's argument instituted a maximalist approach to water use that is a particular mode of thinking which continues to this date. How else can one explain the oxymoron of a drought-stricken island that invests heavily in golf courses, with the sole goal being the short-term economic gains of tourism development, failing, at the same time, to consider the longer-term environmental cost of that investment?[4]

If Thorp's vigorous argument on why it was important to increase water resources was based on a poor understanding of local conditions, then his argument on the how was more practical: Water resources would be increased through surveys and dam-building. Thorp outlined a double strategy for water management: first, surveying the existing groundwater resources and evaluating how to protect them from depletion, excessive drilling and overuse; and second, storing surface water in dams to block water runoff to the sea. This simultaneous emphasis on the survey and conservation of water was in tune with the UN's techno-scientific logic of "taming" (or appropriating, a word that seems appropriate in hindsight), and conserving the "resources" of nature in the name of development. Conservation, since the 1940s, meant the maximisation of development (see, for example, Tucker, 2010: 143). Of course, the combination of survey-and-conservation has this added irony: as soon as unknown water resources became known through surveys, these resources were placed under threat. Effectively, survey and conservation led directly to consumption.

There was a second irony. As a direct result of the Thorp report that prioritised an irrigation-intensive agriculture over other sectors of the economy, the country became even more dependent on water, especially because the key agricultural areas were situated on the plains, where little water was available. Irrigation-intensive agriculture would inevitably require extensive infrastructure, not only for storing water but also for transferring it over long distances. According to this logic, dams and irrigation systems were obviously necessary.

But *who* could increase the island's water resources, according to Thorp? Critiquing previous colonial water practices for limiting their actions to small-scale, isolated schemes (Thorp, 1961: 8), the report declared the new government was the only agent that could overcome the problems of past mismanagement. The island's water resources were so 'exhausted,' and water scarcity was so severe, the argument went, that only the State could oversee the problem for the entire island and could assume the responsibility for 'the development and use of water as fully as may be required' (Thorp, 1961: 8). Besides, centralised state control, Thorp continued, was particularly important in Cyprus not only because of its water scarcity, but also because of 'the complex and fragmented land ownership' and the 'extensive water rights legally vested to private interests,' both of which created obstacles to the development of water resources (1961: 9–14). The report underlined that centralised planning on national and regional scales seemed to be vital particularly for water, because of the integrated nature of the island's hydrological system, where the changing of one part affects the entire balance (Thorp, 1961: 13). The State was to ensure that water is regarded as a 'national resource to be used in the national interest' (Thorp, 1961: 8). Clearly, Thorp was talking about 'the national interest' and 'all the interests of the community' as commonly shared and agreed upon terms, appearing unaware of the realities of intercommunal antagonism in Cyprus. A centrally controlled water development programme was proposed for even the smallest water project. In other words, Thorp's report was a rhetoric – and an aesthetic, as James Scott (1998) convincingly demonstrates – of rationalising legal and land patterns, where the State was not only assumed to be in control of the country's most precious natural resource, but it was also assumed that it would coordinate the management and development of this resource.

While controlling water resources within the structures of government was in line with post-war modernisation theories of the welfare state, encouraging such control was also part of the operational logic of the UN, as it allowed its experts to promote comprehensive planning from within a particular locale. As a consequence, water became the privileged vehicle for nation-building processes, through which the State could exert its sovereignty over the country's most precious natural resource; at the same time, State actors, along with UN experts, gained increased visibility throughout the island, contributing to the strong visual narrative being formed in publications. News items and photos abounded of caring agents generously providing their services and knowledge to the voluntarily participating members of the local society: from the upper mountain levels down to cultivated plains, from the largest river basin to the isolated water drill, and from the techno-scientific meetings with irrigation experts to "on the spot" tips for the uninformed farmer. In addition, the focus on rural areas and the agricultural sector catered to the needs of farmers and peasants – an important 'vote bank' (Panayiotopoulos, 1999: 45) in local elections and also an important constituent group for the State's larger development goals.

All in all, Thorp's suggestions to the UN locked the country into a water-driven future, which substantially increased the need to strengthen State institutions so they could cope with such a technologically and economically challenging task. The sheer volume of the Thorp report, with its very detailed outline of the seriousness of the problem, also made scientific and technical assistance from the UN almost indispensable. Specifically, the report projected the UN as an authority in providing solutions to the problem, while it also pre-conditioned the young State to focus on water generally as key to nation-building processes.

The UN's techno-scientific survey and development programme

Thorp's directives to the UN for the conservation and development of water resources was followed through with comprehensive surveys of minerals and groundwater and with dam construction. These surveys were supported by special funds and specialised agencies of the UN, such as the Food and Agricultural Organization (FAO) that coordinated a programme called 'Water Resources Utilization.' In the early 1970s, these technical assistance programmes were active in such great a number in Cyprus that the UN considered the island nation a paradigm of successful cooperation – a '"Laboratory" for research applied to economic development problems of the Near and the Middle East' (UNDP, 1970: 12).

The developmentalist logic of increasing the use of natural resources was combined with a techno-scientific optimism that the 'assembly and analysis of literally all knowledge about water of Cyprus' was entirely possible as long as enough surveys were completed (UNDP, 1967: 7). Turning water into geo-data and mathematical models, these surveys conducted by the UN were expected to lead to full water control and resource management, so that no drop of water was wasted. Systematic planning would, in turn, lead to water development, while national and local overuse was to be monitored and averted by governmental departments and UN experts, so that eventually the problem would be solved. In other words, the dubious notion of a benevolent State judiciously managing water resources was combined with the problematic idea that nature (specifically water) could be understood and managed by experts, in terms of predictable criteria of resource use and overuse.[5]

Perhaps what best captures the techno-scientific logic of the UN in Cyprus was the motto, apparently introduced by the government's Press and Information Office, 'More dams, more water' (PIO, 1967a). The same press release promised that the extensive programme of dam construction, which the UN's dam experts immediately promoted following Thorp's mission, would result in 'high earnings from agriculture' that would 'outweigh the costs of dams.' This promise highlighted the priorities of the UN: high earnings from agriculture at all costs, through the taming and conservation of nature. The goal of the dam programme planned by the UN was to artificially block all natural streams and rivers of the island and to regulate water distribution in order to extend irrigated land. The programme's optimism went so far as to assume that gradually, even the exhausted groundwater sources could be recharged through specially designed dams.

The pitfalls of techno-scientific developmentalist logic and its ties to nation-building and development, which have been recently documented (Swyngedouw, 2015 and 2014; Kaika, 2005; Bozdogan and Akcan, 2013; Bozdogan, 2002; Demirtas, 2013; Bishop, 2013; Pyla, 2008 and 2007), occurred in Cyprus as well. One pitfall was the intercommunal conflict that formed the political backdrop against which this dream of "more dams" unfolded. Another is the government's official role in appropriating this idea of dams, development and nation-building to advance a rhetoric on unity. And a third pitfall is that dams were also appropriated as a way to advance a supranational vision that aspired to transcend local conflict, echoing another geopolitical reality of that Cold War period: the rhetoric of "non-alignment" and its elusive claim of transcending the superpowers' spheres of influence.

Politics and the UN plan for water development

It seems as though the UN was imagining a development limited by neither water nor politics. The reports mentioned above scarcely address the country's diversity of religion or ethnicity. Perhaps, in the post-independence euphoria, Thorp's hope was that the new nation-state being built would be robust and unified. If he had such a hope for a peaceful future, it was soon dashed. In December 1963, a major conflict erupted between the Turkish Cypriots and the Greek Cypriots. The new president, Makarios, who was a Greek Cypriot, and also the

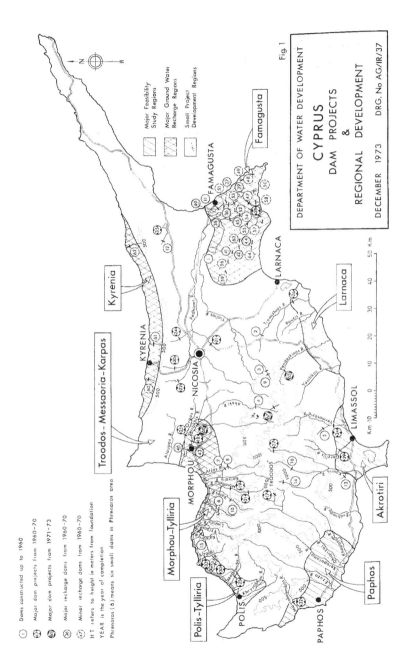

Figure 8.1 Map of various types of dams constructed in Cyprus by 1973 and the regions designated for extended water surveys
(C.A.C. Konteatis, (1973). Dams of Cyprus, Nicosia: Public Information Office for the Water Development Department, Ministry of Agriculture and Resources, Cyprus Republic, p. 2)

Archbishop proposed constitutional amendments to improve the functioning of the State; the Turkish-Cypriot vice-president, Dr. Fazıl Küçük, rejected them. Conflict ensued between the Turkish-Cypriot and Greek-Cypriot populations. By 1964, the Turkish-Cypriot officials had abandoned their government positions and the Turkish-Cypriot population withdrew to enclaves (see, for example, Joseph, 2006; Panteli, 1990). All these violently exposed the underlying lack of trust between the two communities, now manifested in territorial divisions and enclaves that spread throughout the island, making tensions extremely fragmented, localised and unpredictable. The Greek Cypriots challenged the Turkish-Cypriot withdrawal from the State as a unilateral decision and assumed full control of the official government, which continued to receive international aid and funding, especially from the UN. After 1963, the UN combined technical assistance in the country with a peacekeeping mission.

The tensions of 1963–64 created a severe political and state crisis. One might have anticipated that this would change the course of support from the UN, and that the techno-scientific euphoria of "more dams, more water and more development" would transform in response to the intercommunal conflict on the ground. Indeed, the UN did respond by sending a peacekeeping force in 1964. They also offered support to the International Red Cross and assisted people in moving in and out of enclaves; they facilitated communication between the two sides and even helped moving goods and water into enclaves ('UNFICYP helps', 1964; 'Gasoil released', 1964).

Figure 8.2 United Nations military personnel assisting villagers transfer water and goods with a donkey across division lines in Cyprus

(1 April 1973; 52029, UN Photo/Yutaka Nagata, United Nations, New York)

However, other than such peacekeeping gestures, the UN remained fully focussed on their development and water management agenda. One might have thought that schools, civic centres, housing or parks – projects with long and proven associations with social development, the cultivation of civic pride and the like – could have received greater emphasis. Instead, water surveys and dams remained the top priority (UNDP, 1966).

The UN peacekeeping efforts seem to have eased the situation in Cyprus just enough so that the development process could again continue as planned. The reports on UN actions in Cyprus do not discuss the nuances of social justice. Nor did the UN attempt to investigate the roots of this unwelcome conflict; rather, their goal was to avoid derailing development; calling the conflicts 'disturbances' conveniently minimised them. Indeed, as the UN's representative victoriously stated a few years later, 'no single United Nations project was more than briefly interrupted' during the 'disturbances.' As he openly admitted, the UN had clear priorities: first, 'rapid economic and social development,' which would apparently result in the other two UN goals of 'world peace' and 'recognition of human rights' (UNDP, 1966: 6).

The unwavering focus on development continued even after a second conflict broke out in 1967, which resulted in the Turkish-Cypriot community declaring a separate administration (PIO, 1967c). As political talks were initiated between the two communities in 1967–68, the UN became more optimistic about the future of Cyprus, further associating water management projects with other development targets, like industry and tourism (Richmond and Ker-Lindsay, 2001). However, UN optimism denied the complexities of reality. At the grassroots level, water access generated an intense politics of mistrust. The breaking of a pipe or the disruption of water flow was sometimes interpreted as an act of revenge from one side against the other. There were many such incidents throughout the period from 1963 to 1974, which the local press further accentuated by dramatising the antagonism of the two communities and their negative predisposition towards one another.

For example, when the water supply of a Greek-Cypriot village, Exo Metochi, was cut off for two days, the incident was immediately presented as a deliberate act perpetrated by the inhabitants of the neighbouring Turkish-Cypriot village, Epicho ('3.200 water gallons', 1969); some even called it the act of 'Turkish terrorists!' ('Terrorists go too far', 1969). Adding further strain to the drought, the wariness of the two sides often made the management of the problem nearly impossible, especially because water shortage was always a real-life problem that affected economic activities and everyday life. Whenever tensions increased, the two sides would blame each other for problems with the distribution and access to water ('Water lack in Polis', 1965). The blaming started, for example, when the official government passed national legislation to prohibit uncontrolled private drilling for water (PIO, 1964a), and when it proposed a land consolidation law to simplify fragmentary water rights. Such island-wide measures were advanced by the Greek-Cypriot minister of Agriculture and Natural Resources as an effort to 'put an end to the wasteful and arbitrary use of the most valuable production factor' – that is, water (PIO, 1967b). But such measures were also met with distrust by Turkish Cypriots, who accused the Greek-Cypriot side of embarking upon 'a land grabbing exercise' (cited in PIO, 1969). As a consequence, the Greek-Cypriot government officials intensified the implementation of a centralised water management policy, claiming to advance a more "rational" and efficient use of water resources against the allegedly unpredictable behaviour of Turkish Cypriots (PIO, 1973a). In the meantime, the UN continued to play its mediating role, solving pressing humanitarian, economic and environmental problems, and legitimising its presence and its peacekeeping agenda in the process.

Water management was also cited as an example of reconciliatory efforts that transcended the conflict. When a team of Greek-Cypriot and Turkish-Cypriot volunteers took action to resolve water distribution problems in their villages, a leftist newspaper celebrated this as proof that everyday people at the community level were able to overcome the divisive effects of mistrust ('In Pafos villages', 1965). Not only did water prove to be a rather political issue, it was also being appropriated for contradictory political agendas.

Government attempts to forge national unity

The politics of mistrust did not derail the techno-scientific confidence of the UN; in fact, a new ally for the UN emerged from the conflict, namely the nation-state and its central government. If the UN saw its techno-scientific logic as helping to achieve international economic development (and indirectly promoting anti-communism and world peace), the handicapped government, which had lost its Turkish-Cypriot constituents since the 1963–64 conflict, used this same logic for building a rhetoric of a unified and proud nation-state. From their majority perspective, Greek-Cypriot technocrats blamed Turkish-Cypriot advocates of autonomy for their separatism, and accused the enclaves of hindering the entire country's economic development. Appropriating the UN's emphasis on 'techno-economic' development, the official government advanced a rhetoric for island-wide unity, against the Turkish-Cypriot enclaves, which it held responsible for fragmenting an otherwise continuous and unified territory, economy and even natural environment (PIO, 1973a). This fragmentation, the argument went, was becoming an obstacle to comprehensive and integrated planning, which was more efficient and economical, as proven by Thorp's logic.

The State's discourse against territorial, economic and environmental fragmentation thus expanded to identify the Turkish Cypriots and their enclaves as obstacles to the island's development. Against the divisiveness of these obstacles, the official government propagated stories of intercommunal cooperation to shape an argument for social integration and unification. For example, in 1966, the local press spread the message that many 'Turkish' workers were employed on the biggest dam construction site of the time and were working together with 'Greek' workers 'in absolute accord' ('Polemidia Water Dam', 1966). In 1971, when Greek-Cypriot and Turkish-Cypriot villagers in the Famagusta district who had been involved in an irrigation scheme welcomed the director of the UN's FAO, the government announced in an official press release that both communities were benefitting from UN help; it also took the opportunity to argue that addressing livelihood needs could bypass political conflict (PIO, 1971).

The government's emphasis on unity went hand in hand with the techno-scientific logic of water management. It was easy for the government to appropriate the UN rhetoric to its own ends, given how it structured its development plans around the Thorp report. In addition, the government collaborated closely with many UN experts, who were assigned long-term posts within its departments that shaped governmental mechanisms dedicated to water development.

Non-alignment rhetoric and the supranational

The government's efforts to forge national unity were accompanied by efforts to promote its dedication to development within the larger international arena. However, both the claims to national pride and the efforts to insert itself on the map of international development processes were hampered by the intense nationalist antagonism between the Greek-Cypriot and Turkish-Cypriot communities, which resulted in ambiguous allegiances towards the mainlands of Greece and Turkey. The new nation-state was also apprehensive about the presence of the British (which maintained two military bases on the island, even after its withdrawal from the rest of the island), and by extension any interaction with other European colonial powers. Nor could Cyprus define clear allegiances for either of the Cold War polarities, given that both Greece and Turkey (and also Britain) were NATO allies. The multiple ambivalences in Cyprus's geopolitical positioning were further complicated by the United States, who was trying to manage the fragile balances between the two NATO allies by exerting its own economic and political influences in the region. Perhaps it was for all these reasons, as well as the fact that Cyprus's ties to Afro-Asian countries went back to the island's efforts towards self-determination in the 1950s, that the government professed 'a policy of equal friendship with all nations' and opted to join the Non-Aligned Movement (PIO, 1961). The island's president, Makarios, was, in fact, one of the founding members of the movement, and often sought international support from the leaders of Egypt, Yugoslavia and Ghana (Gamal Abdel Nasser, Josip Broz Tito and Kwame Nkrumah respectively) to assert the young State's independence against geopolitical tensions and local threats – especially during times of crisis (see, for example, Hatzivassiliou, 2005).

In August 1964, for example, when the Greek-Cypriot National Guard appeared to be moving into the Turkish-Cypriot Kokkina enclave, Turkish aeroplanes attacked the area. Makarios mobilised his connections with Nasser and Tito so that the two leaders could issue regular press releases in support of the island republic's sovereignty and independence (see PIO, 1964b and 1964c). A few months later, Makarios introduced the "Cyprus problem" at the Second Conference of the Non-Aligned States (5–10 October 1964) in Cairo and hosted Tito in Cyprus immediately afterwards (16 October 1964). In front of a large crowd mobilised by the government, Makarios and Tito jointly stated their commitment to non-alignment agendas for peace and international cooperation. Referring to Yugoslavia's nation-building, Tito particularly emphasised the possibility that different ethnic and religious groups could coexist in the same country, and that this coexistence could be nurtured through cooperation towards development ('Cyprus received and greeted Tito', 1964: 4).

Similarly, in his visit to Ghana in January 1966, Makarios, together with Kwame Nkrumah, also a key leader of the Non-Aligned Movement, flew by helicopter over the Akosombo hydroelectric dam in southeastern Ghana (PIO, 1966), expressing his island government's commitment to Ghana's programme of dam construction. The visit also celebrated the pride emerging from the development of large-scale water infrastructure, a pride that was connected not simply to Cypriot or Ghanaian nationalism, but also to the Non-Aligned Movement's supranational claims, which, to Makarios, offered better alternatives to the local intercommunal conflict.

Furthermore, in 1967, in the midst of a serious outbreak of intercommunal violence, Greek-Cypriot government ministers expressed the desire to join the international conference called 'Water for Peace' in Washington, organised by the United States government. The conference's goal was to map water problems worldwide and to promote experimental technologies. There, the ministers wanted to present Cyprus as a site for a joint endeavour for the United States and the Soviet Union. They planned to propose a joint installation of an 'experimental [desalination] plant' in Cyprus (Cyprus State Archive, 1966). They wanted to turn Cyprus into a laboratory of experimental technologies for exploring solutions to drinking and domestic water problems worldwide, creating along the way a testing ground for technical cooperation that would transcend Cold War antagonisms. These ambitious ideas were part of a larger supranational rhetoric responding to both lingering geopolitical tensions as well as the local conflict within Cyprus. The idea of Cyprus as a hub of global peace was a recurrent theme in the government of the time (see, for example, Phokaides and Pyla, 2012). By turning Cyprus into a hub for an international network of ideas not only about peace but also about people and capital, the government hoped to sidestep complex internal sociopolitical conditions and external dominant geopolitical influences.

Conclusions

What can the history of water management in Cyprus of the 1960s add to interdisciplinary perspectives on water management and its politics? For one, the case of Cyprus highlights the central role of the UN in shaping the techno-scientific agendas of nation-states. That agenda in Cyprus had a dual effect: it locked the country into an economy of increased dependence upon water, and it treated matters of water management and distribution as purely technical ones, divorced from the palpable problems of intercommunal conflict. Furthermore, the case of Cyprus shows how a state's agenda towards forging unity among its citizens went

An island of dams 127

beyond national boundaries; the state evoked supra-political ambitions because these seemed to fit better with the island's geopolitical ambivalences. It was for the purpose of advancing such supra-political aspirations that the government bought into the UN's logic of more and more dams. Contrary to State and UN wishes however, the conflict on the ground served different political ends, for those supporting as well as those opposing the State/UN agendas, making water management a highly contested issue.

What could this history of water politics in the 1960s mean for the prospects of a peaceful and equitable future for Cyprus and its larger region, the explosive Middle East? Water and politics in Cyprus are tightly interconnected in many ways. While water issues are not purely techno-scientific in nature, as we have seen, a solution to the drought problem cannot naively be imagined as also the means to resolve political tensions. Nevertheless, such approaches actually surface these days. Recently, Turkey installed a water pipe on Cyprus, a project inaugurated with the promise to provide a consistent supply of fresh water from Turkey to resolve once and for all the island's water scarcity. This promise obviously overlooked not only the economic and geopolitical impact of the maintenance of such a large-scale technical project but also the social repercussions from the transfer and distribution of water on the island (Hacaoglu, 2015; Tremblay, 2015). Similar ideas that solutions to water problems can transcend conflict have been argued by technocrats as well. In response to the intensified efforts to solve the "Cyprus problem" in recent years, several technical analyses of water management and desalination argue that tackling drought problems can also 'initiate . . . conflict resolution' (Share Water Cyprus, 2012: 5). For example, the 'Share Water Cyprus' study, conducted by a Spanish team, argues that the severe environmental problem of drought highlights the problems caused by the division of the island, and that the urgent and practical need for commonly managing the scarce commodity of water can, in turn, instigate collaboration between the two sides, on other fronts. Such partnerships for water management, according to its proponents, could provide new perspectives for other peace-building efforts as well. For all their hopeful propositions, however, such optimistic conceptualisations of water as a single-minded "political" tool for conflict resolution are ahistorical, much like the purely technocratic view that unequivocally separates water management from any politics. They fail to recognise the complexity of hydropolitics.

Notes

1 Water tanks are a response to Cyprus's irregular annual rainfall levels showing an average of 541 mm for the years 1901 to 1970 and 470 mm between 1971 and 2015. Some of the worst years were 1972–73, 212 mm; 1989–90, 282 mm; 2003–04, 272 mm; 2013–14, 315 mm.

2 The emphasis given by Thorp's mission on water management was also reflected in the report's overall structure and the extended analyses of water issues: the thirteen-page chapter titled 'Water' was placed third (of thirteen chapters), right after the introductory chapters on the Cyprus economy and the guidelines on its future development, while four

128 *Panayiota Pyla and Petros Phokaides*

annexes (of eight) focused on specific technical and management issues: 'Suggested Areas for Geological and Geophysical Surveys for Water'; 'Suggestions for Drilling Programme for Water'; 'Observations Concerning Water Problems by Area'; 'The Water Development Department' (Thorp, 1961).

3 The inadequacy of dams as a permanent solution to water shortage and especially drinking water problems was first noted in 1973. At the time, Cyprus had faced its most severe drought, and the government publicly announced its plans to proceed with a desalination plant (PIO, 1973b). Action in this direction was temporarily delayed because of the high costs of desalination technology, and the government continued the dam construction programme. In the early 1990s water storage volume in dams multiplied, reaching 300 million m³, compared to 50 million m³ in the 1970s and early 1980s. Despite the enormous water storage volume achieved, consecutive series of droughts during the years 1989, 1990, 1993 and 1995 to 1997 signaled the definitive shift towards desalination that has been a major source of drinking water since the late 1990s (WDD, 1999).

4 The first golf course was constructed in 1994 – even as the 1990s presented Cyprus with some of the most severe droughts in its history; many such golf courses followed, as a means to increase tourist influx on the island.

5 Environmental history has shed great light on the flows of such conceptions of natural balance (Worster, 1993; Cronon, 1996). For a perspective on the physical environment specifically, see also Pyla (2012).

References

'3.200 water gallons transferred by Nicosia's Fire Department to the inhabitants of Exo Metochi. The village's water supply was cut by the neighboring Tukrish village Epicho', *Agon* (07 October 1969).

Bishop, E. (2013) 'Control room: Visible and concealed spaces of the Aswan High Dam', in Pyla, P. I. (ed.) *Landscapes of development: The impact of modernization discourses on the physical environment of the Eastern Mediterranean*, Cambridge, MA: Aga Khan Program, Harvard University Graduate School of Design, pp. 72–90.

Bozdogan, S. (2002) *Modernism and nation building: Turkish architectural culture in the early republic*, Seattle: University of Washington Press.

Bozdogan, S. and Akcan, E. (2013) *Turkey: Modern architectures in history*, London: Reaktion Books.

Cronon, W. (ed.) (1996) *Uncommon ground: Rethinking the human place in nature*, New York: W. W. Norton & Company.

'Cyprus received and greeted Tito, the President of Yugoslavia', *Eleftheria* (17 October 1964).

Cyprus State Archive (1966) Letter of Renos Solomides, Minister of Finance to Spyros Kyprianou, Minister of Foreign Affairs, 66/1966/1, 23 August 1966, Nicosia: Author.

Demirtas, A. (2013) 'Rowing boats in the reservoir: Infrastructure as transplanted seascape', in Pyla, P. I. (ed.), *Landscapes of development: The impact of modernization discourses on the physical environment of the Eastern Mediterranean*, Cambridge, MA: Aga Khan program, Harvard University Graduate School of Design, pp. 16–36.

'Gasoil released for Turkish Cypriot ploughing', *The Blue Beret* (3 November 1964).

Hacaoglu, S. (2015) *Turkey finishes longest undersea water pipe to aid North Cyprus,* 7 August 2015 [Online]. Available: www.bloomberg.com/news/articles/2015-08-07/turkey-finishes-longest-undersea-water-pipe-to-aid-north-cyprus [accessed on 20 October 2016].

Hatzivassiliou, E. (2005) 'Cyprus at the crossroads, 1959–63', *European History Quarterly*, 35, no. 4, pp. 523–540.

'In Pafos villages: Turks and Greeks tackle common irrigation problems', *Haravgi* (24 February 1965).

Joseph, J. S. (2006) 'The London and Zurich agreements', in Faustmann, H. and Peristianis, N. (eds.) *Britain in Cyprus: Colonialism and post-colonialism 1878–2006*, Mannheim: Bibliopolis.

Kaika, M. (2005) *City of flows: Modernity, nature, and the city*, New York: Routledge.

Ker-Lindsay, J. (2011) *The Cyprus problem: What everyone needs to know*, New York: Oxford University Press.

Kotsila, P. (2010) *The socio-environmental history of water development and management in the Republic of Cyprus*, Joint European Master Studies on the Environment thesis, Barcelona: Universitat Autonoma de Barcelona.

Nicolet, C. (2002) 'The development of US plans for the resolution of the Cyprus conflict in 1964: "The limits of American power"', *Cold War History*, 3, no. 1, pp. 95–126.

Panayiotopoulos, P. (1999) 'The emergent post-colonial state in Cyprus', *Journal of Commonwealth & Comparative Politics*, 37, no. 1, pp. 31–55.

Panteli, S. (1990) *The making of modern Cyprus: From obscurity to statehood*, New Barnet: Interworld.

Phokaides, P. and Pyla, P. (2012) 'Peripheral hubs and alternative modernisations: Designing for peace and tourism in postcolonial Cyprus', in Heynen, H. and Gosseye, J. (eds.) *Proceedings of the 2nd international meeting of the European architectural history network*, Brussels: EAHN, pp. 442–445.

'Polemidia Water Dam', *Paratiritis* (5 February 1966).

Press and Information Office (PIO) (1961) 'Address of the President of the Republic Archbishop Makarios to the House of Representatives on the 21st August', 21 August 1961, Nicosia: Author.

———— (1962) 'Statement for the press by Dr. C. Hald United Nations representative in Cyprus', 23 December 1962, Nicosia: Author.

———— (1964a) 'Water resources in certain areas to be controlled', 16 May 1964, Nicosia: Author.

———— (1964b) 'President Nasser to President Makarios', 11 August 1964, Nicosia: Author.

———— (1964c) 'Announcement from the Presidential Palace', 13 August 1964, Nicosia: Author.

———— (1966) 'Archbishop's visit to Ghana', 21 January 1966, Nicosia: Author.

———— (1967a) 'UN Report on water development in Cyprus, maximum use of water Resources, as aid in economic development, sought by government with the help of United Nations family', 7 February 1967, Nicosia: Author.

———— (1967b) 'Speech by the Minister of Agriculture and Natural Resources today at the Ledra Palace Hotel during a luncheon given by the association for international development', 15 November 1967, Nicosia: Author.

———— (1967c) 'Re-organization of the Turkish Cypriot administrative system', 29 December 1967, Nicosia: Author.

———— (1969) 'Statement by a government spokesman on the Land Consolidation Law', 10 April 1969, Nicosia: Author.

———— (1971) 'Mr. Aquino tours Famagusta villages', 19 May 1971, Nicosia: Author.

———— (1973a) 'Unfounded Turkish allegations', 16 March 1973, Nicosia: Author.

———— (1973b) 'Meeting for evaluating the impact of drought', 4 April 1973, Nicosia: Author.

Pyla, P. (2007) 'Hassan Fathy revisited: Postwar discourses of science, development, and vernacular architecture', *Journal of Architectural Education*, 60, no. 3, pp. 28–39.

———— (2008) 'Nation-building in Baghdad, 1958', in Isenstadt, S. and Rizvi, K. (eds.) *Modernism and the Middle East: Architecture and politics in the 20th century*, Seattle: University of Washington Press.

———— (2012) 'Beyond smooth talk: Oxymorons, ambivalences, and other current realities of sustainability', *Design and Culture*, 4, no. 3, pp. 273–278.

Richmond, O. P. and Ker-Lindsay, J. (2001) *The work of the UN in Cyprus: Promoting peace and development*, Basingstoke: Palgrave.

Scott, J. C. (1998) *Seeing like a state: How certain schemes to improve the human condition have failed*, New Haven: Yale University Press.

Share Water Cyprus (2012) *Sharing water and environmental values: Peace construction efforts in Cyprus*, Final report of the research project funded by funded by the Catalan Institute for Peace (ICIP), 2011–2012.

Swyngedouw, E. (2014) '"Not a drop of water . . .": State, modernity and the production of nature in Spain, 1898–2010', *Environment and History*, 20, no. 1, pp. 67–92.

——— (2015) *Liquid power: Contested hydro-modernities in twentieth-century Spain*, Cambridge, MA: MIT Press.

'Terrorists go too far: Turkish from Epicho village cut water supply of Exo Metochi. The "pasha" did not allow UNFICYP to visit the village', *Mahi* (5 October 1969).

Thorp, W. L. (1951) 'Some basic policy issues in economic development', *The American Economic Review*, 41, no. 2, pp. 407–417.

——— (1961) *Cyprus – Suggestions for a development programme*, New York: UN.

Tremblay, P. (2015) *Turkey's peace pipe to Cyprus*, [Online], *Al Monitor*, 29 October 2015. Available: www.al-monitor.com/pulse/originals/2015/10/turkey-cyprus-water-pipeline-delivers-fears.html, [20 Oct 2016].

Tucker, R. P. (2010) 'Containing communism by impounding rivers: American strategic interests and the global spread of high dams in the early Cold War', in McNeill, J. R. and Unger, C. R. (eds.) *Environmental histories of the Cold War*, Cambridge: Cambridge University Press.

'UNFICYP helps to ease economic restrictions', *The Blue Beret* (25 August 1964).

United Nations Development Programme (UNDP) (1966) *Five years of United Nations technical assistance in Cyprus, 1961–66.* Nicosia, Cyprus: Author.

——— (1967) *United Nations technical assistance in Cyprus, June 1966 – September 1967*, Nicosia, Cyprus: Author.

——— (1970) *United Nations technical cooperation in Cyprus 1969–1970, 25th Anniversary*, Nicosia, Cyprus: Author.

Water Development Department (WDD) (1999) *Water refineries and desalination plants*, Ministry of Agriculture, Natural Resources and Environment, Nicosia: Republic of Cyprus.

'Water lack in Polis Chrysochou: Water tanks must be transferred away from Turkish neighborhood. Turks irrigate their gardens while Greeks are thirsty', *Haravgi* (21 September 1965).

'Willard L. Thorp' (n.d.) Biographical Notes, [Online], Willard L. (AC 1920) and Clarice Brows Thorp Papers, Amherst College Archives and Special Collections. Available: http://asteria.fivecolleges.edu/findaids/amherst/ma220_bioghist.html [20 October 2016].

Woodhouse, C. M. (1958) 'The Cyprus problem', in Philip W. T. (ed.) *Tensions in the Middle East*, Baltimore: The Johns Hopkins Press.

Worster, D. (1993) 'The shaky ground of sustainability', in Sachs W. (ed.) *Global ecology: A new arena of political conflict*, London: Zed Books.

9 Counter-infrastructure as resistance in the hydrosocial territory of the occupied Golan Heights

Muna Dajani and Michael Mason

Introduction

This chapter examines the development of water infrastructure by indigenous Arab communities as a political response to hydraulic domination by an external occupying power. The Israeli military occupation, since 1967, of two-thirds of the Syrian Golan Heights, followed in 1981 by de facto annexation, created a situation where only five Syrian Arab villages, clustered in the north, remain with access to 20,000 dunums (2000 ha) of cultivated land, compared to 80,000 dunums (8000 ha) of cultivated land farmed by Jewish-Israeli settlers (Keary, 2013). Israel, as occupying power, has transformed water infrastructure in the Golan Heights, constructing artificial lakes, dams and reservoirs to harness water for settlement agriculture. Such actions have severely restricted the agricultural practices and water management schemes of the Syrian (mainly Druze) farmers of the Golan. These farmers have responded with a counter-hegemonic water infrastructure and associated land use choices designed to bypass discriminatory restrictions on the abstraction, storage and use of water for agriculture.

Using settler colonial theory and the concept of hydrosocial territories, we examine the production and effects of this insurgent infrastructure. Settler colonial theory offers explanatory propositions on the exceptional governance accompanying the coercive takeover and control of a foreign territory, typically expressed as the permanent appropriation of land and other natural resources, alongside the political and economic subordination of the indigenous population (Veracini, 2013; Veracini, 2010; Wolfe, 2006). While this theory highlights the dynamic of de-territorialisation driving settler colonialism, which can plausibly be attached to Israeli hydraulic and settlement practices in the occupied Golan Heights, we also draw on the concept of hydrosocial territories (Boelens et al., 2016) to capture the diverse material and symbolic interactions at play where water infrastructure is a conflictual site of state power. As discussed below, the Israeli hydraulic mission for the occupied Golan Heights manifests divergent experiences and imaginaries for the state and its settler subjects and for the indigenous local Syrian ('Jawlani') population contesting control over water flows.

We first discuss settler colonialism as predicated on a dual logic of the de-territorialisation of the indigenous populations whilst at the same time advancing the territorialisation of a settler colonial settler society. In the occupied Golan Heights, the configuration of hydrosocial domination is enacted by state appropriation of land and water resources, providing settlers with continuous, subsidised and connected water infrastructure, whilst systematically denying equal water access to the non-settler indigenous population. For the Jawlani, the construction of what we refer to as counter-infrastructure creates hydrosocial space both for autonomous water access and the forging of a community solidarity resistant to forced settler colonial assimilation. The making of alternative hydrosocial realities, fostered by communal water management norms, creates complex socio-natural interlinkages where, for the Jawlani, pragmatic coexistence with the occupying power sits alongside dissenting hydrological flows. Hydrosocial territories may overlap across shared terrain and watersheds, but in the occupied Golan Heights, where the de facto sovereign has attempted to institute a monopoly of material and symbolic control over water resources, the water counter-infrastructure of the Jawlani exists precariously as a defensive, but non-violent, manifestation of collective self-governance.

Setter colonialism as de-territorialisation

Veracini (2013) defines settler colonialism as a political-economic formation which aims to expel a majority of the indigenous population, creating a new territory for the settlers of the conquering state. Settler colonialism deploys an expansionist spatial logic of accumulation by dispossession. As Lloyd notes:

> what distinguishes a settler colony from an administrative or extractive one is in the first place the settlers' focus on the permanent appropriation of land rather than the political and economic subordination of the indigenous population, the monopolisation of its resources, or the control of its markets.
>
> (2012: 66)

As a land-centred endeavour, and unlike the fixation of historical colonialism on natives as labour, the Zionist settler colonial project in Palestine focused on 'Jewish land' and 'Jewish labourers' and the control of natural resources. The establishment of agricultural cooperatives, kibbutzim, was the epitome of that vision during the early years of the State of Israel, creating a new society devoid of the native and harnessing natural resources for the benefit of national development.

The transformation of the inherent meaning in land, territory and terrain has been archetypal of state and settler interventions (Kolers, 2009). Settler societies do not recognise indigenous conceptions of land and property, typically viewing indigenous populations as failing to make productive use of land. The strategy of "indigenous de-territorialization" (Choi, 2016: 13) has been deployed to strip from these populations both from land and other resource entitlements and also affective connections to the land. A settler project succeeds, therefore, when it cements its control of the land it occupies, establishes sovereign political authority and completes its termination of autonomous indigenous forms

(Veracini, 2013). Under such conditions, territorial power is associated with fear and violence, where the settler/indigene distinction reproduces binaries of inclusion/exclusion and friend/enemy (Elden, 2010; Lefebvre, 1991: 280).

Under the land-centred ambitions of settler colonialism, de- and re-territorialisation around water is central to the re-configuration of socio-natural conditions. Hydrosocial territories transcend their biophysical elements (as expressed by the hydrological cycle) to reveal socio-environmental values, meanings and imaginaries of water mixing with social, economic and cultural dynamics of power (Zwarteveen and Boelens, 2014; Swyngedouw, 2009; Swyngedouw, 1999; Wittfogel, 1957). Exploring the spaces and conduits through which water flows therefore requires exploring both physical and social processes of hydrological circulation. Viewing the occupied Golan Heights through a hydrosocial territory lens can strengthen our understanding of how water dynamics are embedded in political and socio-natural conditions. Hydrosocial territories are "socially, naturally and politically constituted spaces that are (re)created through the interactions amongst human practices, water flows, hydraulic technologies, biophysical elements, socio-economic structures and cultural-political institutions" (Boelens et al., 2016: 1). This means that hydrosocial territories exist in pluralist forms, articulating contrasting notions of what such territories mean and to whom, and often featuring negotiations and struggle over the governance of water resources (Hoogesteger et al., 2016). The relevance of territorial struggles, in the context of settler colonialism, is that they "entwine battles over natural resources with struggles over meaning, norms, knowledge decision-making authority, and discourses" (Boelens et al., 2016: 8). The hydrosocial configuration of the Israeli hydraulic mission, which structures water allocation and use in occupied Palestinian and Syrian territories, reproduces stark asymmetries in water availability between settler subjects and the indigenous Arab populations. The state re-territorialises new hydrosocial territories through coercive (sovereign) power and the imposition of regulatory authority, but can assimilate indigenous communities by connecting them to settler water infrastructure and networks. Hydrosocial territories become (re)configured through infrastructure, entailing major impacts for different water users' identification with the physical environments while simultaneously altering political order and establishing a hydro-political network hierarchy (Duarte-Abadia et al., 2015). Moreover, new technologies and infrastructure to govern water also constitute environmental subjects: users, managers, technicians (Agrawal, 2005).

While space and scale are important with a hydrosocial analytical lens, so is "autonomy and self-determination" over socio-natural spaces and their symbolic meanings (Hoogesteger et al., 2016: 102). As revealed below for the occupied Golan Heights, not only is such a lens relevant for how the state orders and governmentalises water, but also how counter-strategies over water allocation and management are practiced by indigenous Arab communities, playing out in the contested configuration of hydrosocial territories. While state policies promote infrastructure as an aspiration for development and modernity, and as a tool for social ordering and governmentality (Zwarteveen and Boelens, 2014), water infrastructure acquires physical form through the concretising and channelling of state power. Through infrastructure, the state can distribute material capital (of water flows) but also social, ideological and symbolic capital (Swyngedouw, 2004). Hence, state infrastructure can become sites of structural violence and used as tools of exclusion and marginalisation (Ibrahim, 2017; Rodgers and O'Neill, 2012; Anand, 2011). In relational terms, infrastructure is both social and ecological: "it means different things to different groups and it is part of the balance of action, tools and the built environment, inseparable from them" (Star, 1999: 377). The Golan Heights, as a case of settler colonial conquest, produces water infrastructure as a duality – a story of infrastructural development for the settler and settler state interests at the same time as de-development of indigenous infrastructure (Ibrahim, 2017). The settler state is expansionist and ethnocentric: in the case of Israel, water infrastructure has been a core technology for re-territorialisation in the occupied Arab territories, establishing centralised control over water and land resources for the purposes of settler colonisation, whilst at the same time rendering as illegal and/or marginal the local water infrastructure of the indigenous Arab populations (Weizman, 2007; Yiftachel, 2006).

Settler colonialism and the Israeli hydraulic mission

With such strong self-representations about the 'permanence' of settler colonial projects (Choi, 2016), it is not surprising that settler colonial infrastructures are cemented and 'concretised' (Meehan, 2014), imprinted onto the landscape as markers of a new ethno-geography. In Eric Zakim's *To build and be built: Landscape, literature and the construction of Zionist Identity*, titled after the lyrics of a Zionist folk song – "we came to the land to build and be built by/in it" – he refers to the "transformation of Palestine from an inimical environment into a quintessentially Jewish space" (2006: 1). The production of Jewish space was viewed as a modernist, future-looking and transformative endeavour. In the creation of the Israeli state, Zionism was portrayed in large part as a hydraulic mission, constructing mega-structures for modernising water supply, bolstering state and nation-building efforts and realising dreams of irrigating the desert. With technology, raw materials and expertise, the Zionist and Israeli regimes succeeded in their quest to develop the Jordan River Basin and thus realise the Israeli state hydraulic bureaucracy.

Theodor Herzl, the founding father of Zionism, regarded hydroelectric power as the economic basis of the new society in Palestine (Smith, 1993: 118). Water infrastructure development, especially to increase water availability for further economic development, continued to be a building block of the Israeli mindset for maintaining and enhancing a rational and technologically advanced society (Feitelson and Rosenthal, 2012). The Zionist/Israeli hydrosocial infrastructure begins prior to the establishment of the state, with its construction as a nation-building endeavour in the 1930s; what Feitelson and Rosenthal (2012: 273) call the "Zionist hydraulic mission" era. The Rutenberg concession, granted by the British Mandate to the Zionist pioneer Pinhas Rutenberg in the 1920s is case

Counter-infrastructure as resistance   135

in point of the Zionist aspiration to tame nature and develop a new society. Rutenberg, with exclusive rights to exploit the waters of the Jordan, constructed a hydropower plant on the banks of the River Jordan, acquiring large amounts of land on both banks of the river and controlling the confluence of the Yarmouk and Jordan Rivers. Until its destruction and halting of its operations in the 1948 war, it exemplified the Zionist imaginary of hydrological domination and control (Meiton, 2015). The 1950s marked a period of US-led negotiations over the allocations of the Jordan River Basin water between the newly established nation states (of Israel, Jordan, Syria and Lebanon) in the wake of the 1948 war, which left over 700,000 Palestinians expelled and dispersed around the region.

The failure to reach a regional agreement on water strengthened these newly established states to assert their territorial sovereignty over the contested water of the Jordan, resulting in unilateral hydraulic projects that transformed the watershed. To reinforce its hegemonic hydraulic mission, between 1953 and 1964 Israel constructed the National Water Carrier, its biggest infrastructure project and the epitome of water infrastructure as a technology for nation-building. Diverting 350 million cubic metres annually from the Jordan River Basin towards the coastal cities and reaching Al Naqab (Negev), and causing long-term environmental deterioration to the lower Jordan River, this infrastructure project also provoked Jordan and Syria to intensify unilateral water withdrawals from the same river basin. The National Water Carrier has been labelled "a centralised technical apparatus through which all the water of the state was regulated" (Harris and Alatout, 2010: 153). Following its occupation of the West Bank, Gaza Strip and the Syrian Golan Heights, Israel further secured control over the headwaters of the Jordan River, in addition to controlling the groundwater of the West Bank (Zeitoun, 2008; Zeitoun et al., 2012). Israel's realisation of its hydraulic mission attests to the claim of Zeitoun and Warner (2006: 445) that "a state with the ability to plan, construct and operate large infrastructure projects has the physical ability to change the hydrogeology of the resource, thereby creating new hydro-strategic and hydro-political realities." Through an assemblage of dams, pipelines, pumping stations, reservoirs and diversions, the development by Israel of a centralised water infrastructure, supported by institutional arrangements and engineering expertise, facilitated the emergence of a hydraulic bureaucracy. These and other infrastructures of territorialisation have been "central to the symbolic and material geographies of Zionism" (Salamanca et al., 2012).

The Israeli water law of 1959 supported this project, constructing water as public property under the centralised authority of the state, justified as necessary to nation-building and the exercise of Jewish sovereignty under conditions of perceived water scarcity (Alatout, 2007). These categories and their articulation with one another made the water law a core instrument in Israeli structures of power, solidifying the Jewish character of state institutions and contributing to the marginalisation of Israel's Arab/Palestinian citizens. Lipchin claims that the Zionist ideologies driving water and agricultural policy in Israel have "left a legacy of mismanagement and environmental degradation" (2007: 251). Selection of water projects in the newly established state was not based on economic logic, but rather on achievements to create a cultivated and development society, with agricultural production and irrigation of the desert a national priority. The growth of the Israeli water sector facilitated, through infrastructure development, the "emergence of a central structure designed to carry out ideological directives" (Galnoor, 1978: 349). Mekorot, the Israeli water company, founded in 1937 before the establishment of the state, was owned jointly by the government, Jewish Agency and the National Federation of Labour 'Histadrut' (Bilski et al., 1980). Today, Mekorot is a wholly owned government company, under the Ministry of Energy and Water and the Ministry of Finance. It is formed of a group of companies which control, plan and manage 100 water mega-projects throughout Israel and occupied territories. Also central to the historical development of Israel's water supply, through land acquisition and control over water sources, has been Keren Kayemeth LeIsrael – Jewish National Fund (KKL-JNF). Indeed, KKL-JNF has increased Israel's water supply to such an extent that it has been dubbed "Israel's fourth aquifer" (JNF, 2017).

With the territorialisation of water governance by the Israeli state in occupied territories, space is (re)produced by authoritative and ideologically-driven water governors, framing and dictating a hegemonic hydrosocial territory. It also shapes disenfranchised communities as water users facing state-led institutional and technological discrimination. These indigenous water users and communities experience drastic alterations to their socio-natural conditions of life. Both the occupied Palestinian and Syrian territories there have seen, in response, differentiated modes of organising and political mobilisation to protect residual hydrosocial territories (Boelens et al., 2016; Agrawal, 2005). Due to its selective and systematic discrimination, the Israeli hydrosocial territory disciplines and reinforces structural violence in the form of infrastructural and territorial violence, providing legitimacy to its hydro-hegemony of water use, control and access. By examining water infrastructure and its practices on a localised level, the following section aims to re-politicise water use practices and explore counter-infrastructures and reconfigurations of local hydrosocial territories in the occupied Golan Heights.

State infrastructure and hydrosocial domination in the occupied Golan Heights

The Golan Heights presents a set of physical and geopolitical strategic characteristics, rich in water resources and consisting of a very fertile plateau, conditions which justified and prioritised its capture and control (Ibrahim, 2017; Ram, 2015). More than 147,000 people distributed in 163 villages and towns in addition to 108 farms (Mara'i and Halabi, 1992) lived in the Golan Heights prior to 1967, creating a thriving community of diverse ethnic and religious backgrounds under the Syrian state. Before the occupation, no real exploration of the groundwater was taking place in the Golan Heights, except for a few shallow wells. This was due to the lack of capital to drill such wells and the rain-fed character of agricultural produce. In the Golan Heights under Syrian sovereign control, crops such as almonds, figs and grapes were the most common.

Following the 1967 war, Israel disposed a thriving population of villages and towns in the Quneitra province in the Golan Heights, displacing most Arab residents from the land, with only five villages remaining intact and a population of 6,000 only remaining in the newly occupied territory – Majdal Shams, Ein Qinya, Mas'ada, Buq'atha and Al Ghajar. The forcible transfer of native inhabitants was followed by the systematic destruction by the Israeli military of Arab villages and farms, facilitating the land appropriation, settlement building and the transfer of Israeli settlers into the region – all breaches of international humanitarian law (Murphy and Gannon, 2008: 147–152). With this abrupt transformation, Syrian territorial sovereignty was replaced by Israeli military control. A series of military ordinances were issued by the Israeli military commanders, covering all aspects of civilian life, including land ownership and use, freedom of movement and political expression, right to demonstrate and economic activities. In regards to water, the military commander established a permit system for drilling of new wells, regulated pumping quotas and most importantly dictated how water disputes are resolved. Military Order 291, issued in 1968, declared all pre-1967 land and water-related arrangements as invalid (Amnesty International, 2009), deeply impacting and distorting communal and local customary law deployed for decades by the local Arab population. Military Order 120, issued in 1968, gave Israel the full rights to manage and oversee the water resources in the area, allowing military access to any area containing water works, even forcing the local population to fully cooperate with the military forces in identifying local water resources and sharing details regarding their quantity and quality (Keary, 2013).

In 1982 the Israeli Knesset passed the Golan Law, annexing the occupied Golan into its territory, a move considered illegal under international law (Ó Cuinn, 2011). This ultimately saw the enactment of Israeli civil law in all aspects of Jawlani life, although Israeli efforts to achieve political assimilation of the Syrian Arab population – through forced Israeli citizenship – were largely resisted. The Israeli government eventually issued Israeli travel documents stating "undefined/unknown nationality" as a punitive action for refusal of Israeli citizenship – a denial of "ethnic and national specificity" consistent with settler colonial governance (Veracini, 2006: 19). The move from military control to civil administration preserved a central role for the state consistent with Israeli water policymaking. The Water Law of 1959, formally enacted across the occupied Golan Heights by annexation, treats water sources – springs, streams, rivers, lakes and other currents and water reservoirs – as public property controlled by the state to fulfil the needs of the population and national development. State infrastructure was developed by an Israeli water management company operating specifically in the Golan, called Mey Golan ('Water of the Golan'). Funded by KKL-JNF, Mey Golan has played a pivotal role in re-configuring the territorial arrangements of water governance in the region, establishing themselves and Mekorot as water governors serving the political and economic interests of the Jewish settlements, including irrigation needs for the expansion of agricultural production. Since the creation of Mey Golan in 1978, 16 flood and stream dams were constructed in the northern part of the Golan, halting the flow to natural springs and waterways which the local Jawlani depended on. In addition, eight deep wells were also drilled exclusively pumping water to settler colonies to feed their agricultural developments (Keary, 2013). Mey Golan has developed the infrastructure of the occupied Golan by building 14 stream reservoirs throughout the region, collecting stream and flood flow, in addition to the drilling of several deep wells, to capture approximately 35 million cubic metres every year (Author's interview, 2016). Lake Ram, a natural volcanic lake long used by the local population for livestock and minor agricultural purposes, has been appropriated by Israel, controlling access to water from the lake. A pumping and distribution facility is located next to the lake, pumping water to meet the water demands of the Israeli settlers at a subsidised cost.

The water infrastructure investments by Mey Golan reinforce the Israeli hydraulic mission of establishing hydrosocial dominance over the Jordan Basin waters, with the National Water Carrier acting as the core conduit. In the occupied Golan Heights this furthers the de-territorialisation of the indigenous Jawlani communities, not only through limiting and prohibiting their access and use of water resources, but also by disregarding their customary water management practices and attachments to land. Systemic discrimination over water allocation and access to Israeli water infrastructure is merely one set of resource-based inequalities experienced by the Druze inhabitants of the Golan Heights. Syrian farmers are blocked from acquiring new land and face heavy restrictions on water usage, paying higher rates for piped water than settlers, who benefit from subsidised water distribution (Wessels, 2015; Ó Cuinn, 2011: 92–93). For example, Jawlani farmers face more restrictive access and have to pay higher prices to access water from Lake Ram; they are also impeded by Mekerot requirements to invest in their own pumps and pipes to extract water from the lake. It was the Israeli imposition, since the 1970s, of strict controls on water abstraction and storage by Arab farmers that initially triggered their political mobilisation around the issue of water rights, including the right to use, access and store water.

The next section examines the development of water counter-infrastructure (Meehan, 2014; Swyngedouw, 2004) by the Jawlani as a collective political project to resist the hydrosocial territorial dispossession under settler colonialism in the occupied Golan Heights. In her broad theorisation of water infrastructure, Meehan looks at local practices striving for water autonomy, like the rain barrels and tubewells used by water users in defiance of the legality of state control of resources and as a proof of the limitation of the political reach of the state and its centralised infrastructure, however entrenched it is in the landscape (Meehan, 2014: 223; Harris and Alatout, 2010). Counter-infrastructure combines material and symbolic resources to advance a space of autonomous action resisting hydrological domination; under a heavily asymmetrical water governance structure, this non-violent resistance treads a careful, precarious path of political possibility which may be closed off at any time. In the occupied Golan Heights, the Jawlani farmers have generated a subaltern hydrosocial territory that intertwines with the Israeli state-administrated hydrosocial territory in a complex dynamic of coexistence and contestation.

Water counter-infrastructure as resistance

> We have ambitions to reclaim our water, since Mekorot does not own it but rather has possessed it. Gradually we will reduce Mekorot's use of the water and reclaim it for our own use.
>
> (Farmer and political veteran, interview, Majdal Shams, December 2016)

The tactics which led to the development by the Jawlani of counter-infrastructure for irrigation water arise firstly from biophysical setting. The geographical location of the Syrian villages at an elevation between 600 and 1200 metres above sea level, in addition to their position at the headwaters of the Upper Jordan Basin, necessitates agro-development that conforms to these environmental conditions. Such conditions dictate the planting of certain crops, with experience since the 1950s favouring fruit trees, in particular apple orchards, as the most suitable and lucrative crop for these farming communities. Prior to 1967, the Golan Heights villages belonged to the Syrian province of Quneitra, and were populated by predominantly subsistence farmers, who even then enjoyed a large degree of autonomy from the state (Batatu, 1999). Farmers in field interviews conducted in 2013 proudly boasted of historical examples of community mobilisation to manage their natural resources and develop their agriculture. From channelling the spring water of Ein El Tufaha through basic dirt channels, the farmers began flood irrigation to the central agricultural valley of Al Marj to irrigate newly planted apple trees. The significance of this collective infrastructure, built in the 1940s, during the French mandate and prior to the creation of the Syrian state, lies in its formative structuring of a hydrosocial territory, where Arab livelihood practices intertwined with national, religious and geographical identity, forging an enduring ethno-geographic community. For the next three decades, there was a steady transformation of rain-fed subsistence agriculture to flood irrigation, with 6,400 dunums under apple tree cultivation by 1967 (Abu Jabal, 1993). The water resources that were utilised by the farming communities comprised of springs, a few shallow wells and man-made pools constructed by farmers to capture rainwater, in addition to the volcanic natural pool of Birket Ram (Lake Ram).

In the wake of the 1967 occupation and the effective de-territorialisation of the indigenous population of the Golan Heights, the five remaining Arab villages faced enormous pressure to keep their communities intact while also adapting to the new order of the occupation, which included immediate military control over their land and water resources. Driven by their quest to protect and expand their economically profitable apple production, and in light of Israeli land confiscation plans, the need to expand cultivable lands became essential. In light of their shrinking hydrosocial territory, collective efforts were mobilised to protect their remaining property and natural resources. Israel utilised legal powers, deriving originally from an Ottoman land law, allowing the state to seize land uncultivated for three years – powers that have been used extensively to confiscate land from Palestinians (Forman, 2011). To defend the land from this mechanism of state appropriation, the farmers needed to expand their cultivated land, carrying out agricultural reclamation of neighbouring hills. This required collective action and the local utilisation of labour, machinery and skills, initiating an 'agricultural revolution' to protect Arab lands and livelihoods.

Restrictions on water use were intensified in the 1970s by Israeli authorities, who have declared all water sources to belong to the state and any collection and use of water not licensed by the government to be a violation of the relevant military orders. With the confiscation by Israel of Lake Ram, the damming of stream and spring flows and the pumping of ground water solely through the Mekorot and Mey Golan companies, the hydrosocial territory of the occupying power was consolidated, undermining and displacing Jawlani hydrosocial space. However, alternative counter-infrastructures of water were also being constructed by the Jawlani farmers in reaction and opposition to what was termed, in a National Statement issued in 1982 by the Jawlani communities as a global call for solidarity, as "rape of our groundwater", "the theft of Birket Ram" and the denial of an ethno-geographic identity. Infrastructural violence was also referred to in this National Statement, highlighting how Israeli military rule had halted a community water project to transport water to newly reclaimed agricultural land (National Statement, in Al Batheesh, 1987: 39).

The 1980s witnessed a splurge in the construction of small reservoirs to catch rainwater, part of the collective Arab efforts to increase the area under agricultural cultivation to protect the land from confiscation and increase the planting of apple trees to establish facts on the ground. Accounts of water reclamation from Jawlani farmers interviewed in spring 2012 included one where farmers of Masa'da village, where Lake Ram is located, pumped water in the middle of the night to mobile tankers and used the water from the lake to irrigate their crops planted a few hundred metres away. To secure sufficient water sources to irrigate the apple orchards, hundreds of circular metal tanks, with a volume between 300 and 1000 cubic metres were built and erected in agricultural land to collect rainwater. These metal tanks started dotting the Jawlani landscape in the 1980s, in defiance of the Israeli water law which prohibited the harvesting of rainwater for private use. In a cat-and-mouse game with the Israeli authorities, the construction of such tanks by Arab farmers was followed by acts of state demolition, fines and taxation to limit the development of such structures. However, many remain to this day as a reminder of acts of asserting and reclaiming indigenous water rights, although their current use is limited. Today, 12,000 dunums of apple orchards are planted by Jawlani famers, creating a distinctive landscape that is inseparable from its inhabitants' identification with the land of the Golan Heights.

Collective reclamation of water rights by constructing metal tanks as water reservoirs constitutes a tactic of collective action and not merely an act by individual farmers. The continued struggle and mobilisation against water appropriation by the Israeli state also took the form of the establishment of water

Figure 9.1 Shallow pools (left) and rainwater harvesting tanks (right) marked the beginning of a water counter-infrastructure

cooperatives, the first of which were established in the mid-1980s. The Jawlanis began working collectively to reclaim water resources to irrigate their apple orchards. Negotiation with the Israeli water company Mekorot began in the 1990s, creating channels of negotiation and lobbying to acquire water rights and receive quotas for irrigation water from freshwater sources, including Lake Ram. In 2012, 18 cooperatives existed, representing farmers from all the Jawlani villages. Thus far, they have secured over 40 per cent of the water irrigation needs of Arab farmers with apple orchards.

Cooperatives also paved the way to the upscaling of demands and the beginning of a re-configuration of the local hydrosocial territory through the establishment of the right to infrastructure. While Mekorot allowed the selling of the waters of Lake Ram, it was conditional on the Jawlani farmers providing their own means of transporting this water to their lands. Through the cooperatives, the Jawlani farmers began working with local contractors to design the most suitable network of pipelines, pumps and meters to ensure collective administration of this water. Once again, the farmers found themselves leading a re-configuration of their hydrosocial territories that had begun in the 1940s: construction of infrastructure to capture their reclaimed water, achieved through collaborative struggle and mobilisation which now is occurring under conditions of contestation and protracted occupation.

Resembling the water tanks that dot the landscape of the villages, the pipelines now emerged as an upscaled counter-infrastructure, creating facts on the ground which defy and compete with the hegemonic narratives and practices of state infrastructure (including law). The network of pipelines has made permeable the state's unequivocal control over water, as it transports water (the physical, social and symbolic meaning of it) away from state power and control to that of farmers irrigating their apple orchards. Through such an assemblage, the apple orchards area expanded to 12,000 dunums by 2012 and re-configured the state-Jawlani interaction. Farmers were thus able to irrigate a crop, considered "the bloodline of life" in the remaining villages of the Golan and an important signifier of

Figure 9.2 The upscaling of counter-infrastructure through the construction of community-funded pipelines and pumps

community resistance, by building a water counter-infrastructure which is not state-funded, -owned or -operated. This bottom-up infrastructure solidified the communities' belonging and steadfastness on the land and enhanced their bargaining power vis-à-vis the state to acquire larger quotas of water for irrigation.

The role of cooperatives has not only been to lobby for and negotiate water allocation quotas and prices with Mekerot, but also more importantly to begin the construction and development of a viable water irrigation infrastructure. The Israeli state has negotiated with the farmers as 'water consumers' while, at the same time, the farmers' collectively developed their own physical infrastructure, such as pipelines, pumps and irrigation networks. For the Jawlani the construction of this counter-infrastructure creates hydrosocial space both for autonomous water access and the forging of a community solidarity resistant to forced settler colonial assimilation. The making of alternative hydrosocial realities, fostered by communal water management norms, creates complex socio-natural interlinkages where, for the Jawlani, pragmatic coexistence with the occupying power sits alongside dissenting hydrological flows – a material and symbolic network of resistance to the re-configuration and de-territorialisation of their hydrosocial space. That this upscaling of infrastructure was funded and constructed and funded by Arab farming cooperatives reiterates the significance of autonomous water governance as a site of political self-determination and collective identity (Hoogesteger and Verzijl, 2015; Perreault, 2003). The development of the irrigation infrastructure for the local farming communities of the occupied Golan Heights can therefore be seen as an example of a "(semi) autonomous grass-roots hydrosocial territory" (Hoogesteger et al., 2016: 97).

Conclusion

> Mekorot supplying or, rightly put, 'selling' us water, is a type of sedation of our suffering.
>
> (Farmer and political veteran, Interview, Majdal Shams, December 2016)

The contested hydrosocial territory of the occupied Golan Heights reveals how indigenous water rights and uses have been negatively impacted and transformed by settler colonialism, which has re-configured socio-natural relationships by implanting new legal, institutional and technological norms and practices. Israel has used water infrastructure development in the Golan Heights as a technology of state territorialisation, furthering a transformative occupation that is at odds with international humanitarian law. The modernisation of water infrastructure by Israel has been justified by a well-established narrative of hydrological security and control within an unstable, and often hostile, regional context. In its attempt to construct its waterscape, the Israeli state has disregarded the water management practices of the indigenous Arab population, realising a highly centralised and securitised water infrastructure. The assumption has been that the assimilation of the 'native' water use and users could be achieved by taming nature and its 'natural' users. Referring to the Rutenberg hydropower concessions granted by the British mandate to Zionist investors in the 1930s, Meiton observes that "rather than being a party to the agreement, then, the Palestinian Arabs were given a status more like the other natural elements that had to be tamed for the sake of modern development" (Meiton, 2016: 212). Such dismissal of the agency of local populations, in this case the Jawlani in the Golan Heights, is omnipresent in Israeli's water policymaking and infrastructure development. Those Syrian Arabs remaining in the occupied Golan Heights who have not taken Israeli citizenship are similarly part of an indigenous or natural state – unacknowledged background conditions for the palimpsest of Jewish sovereignty and Zionist colonisation. As in Palestine, their political marginalisation accompanies the cultural displacement of a distinctive ethno-geographic community (Ra'ad, 2010).

Although a hydropolitics lens reveals the highly nationalised and securitised socio-nature of the waters of the Golan, including their material and symbolic appropriation under the basin-wide circulatory dominance of Israel, this analytical framing tells only part of the story. On a local level, the occupied Golan Heights remains home to an indigenous agricultural society, comprising today around 20,000 Syrian Arab inhabitants. The material and discursive monopolisation of Golan water by an occupying sovereign has displaced alternative water practices and meanings expressed by Syrian farmers and other Arab water users. These local water practices reflect and react to, but are not reducible to, political changes that have taken place in the relation between the Jawlani and Syria as a national homeland, and also with Israel as the occupying power since 1967.

A counter-infrastructure of water, from rainwater reservoirs to parallel pipe-lines, has been designed by the Jawlani to bypass discriminatory restrictions on the abstraction, storage and use of water for agriculture. Informed by communal water management norms, this counter-infrastructure marks out an alternative hydrosocial space featuring autonomous water use and the forging of a community solidarity resistant to settler colonial assimilation. The extent to which this counter-infrastructure is submissive or insurgent rests on a daily calculation of political opportunities and futures. As with other forms of everyday opposition under asymmetrical relations of power, there is a "massive middle ground, in which conformity is often a self-conscious strategy and resistance, is a carefully hedged affair that avoids all-or-nothing confrontations" (Scott, 1987: 285). In the occupied Golan Heights, where Israel has asserted a monopoly of control over water resources, the water counter-infrastructure of the Jawlani exists pre-cariously as a defensive, non-violent manifestation of collective self-governance.

References

Abu Jabal, H. (1993). *The case of agriculture: Proceeding of the first study day on twenty-five years of Israeli occupation of the Syrian Golan Heights.* Jerusalem: The Arab Association for Development.

Agrawal, A. (2005). *Environmentality: Technologies of government and the making of subjects.* Durham, NC: Duke University Press.

Alatout, S. (2007). State-ing natural resources through law: The codification and articulation of water scarcity and citizenship in Israel. *The Arab World Geographer* 10(1), pp. 16–37.

Al Batheesh, N. (1987). The Golan: Confronting the Occupation 1967–1987. *Al Itihad* [Arabic Source].

Amnesty International. (2009). *Troubled waters – Palestinian denied fair access to water.* www.amnestyusa.org/pdf/mde150272009en.pdf, last accessed September 2016.

Anand, N. (2011). Pressure: The PoliTechnics of water supply in Mumbai. *Cultural Anthropology* 26(4), pp. 542–564.

Batatu, H. (1999). *Syria's peasantry, the descendants of its lesser rural notables, and their politics.* Princeton: Princeton University Press.

Bilski, R., Galnoor, I., Inbar, D., Manor, Y. and Sheffer, G. (eds.) (1980). *Can planning replace politics? The Israeli experience.* The Hague: Martinus Nijhoff Publishers.

Boelens, R., Hoogesteger, J., Swyngedouw, E., Vos, J. and Wester, P. (2016). Hydrosocial territories: A political ecology perspective. *Water International* 41(1), pp. 1–14.

Choi, S. E. (2016). *Decolonization and the French of Algeria.* London: Palgrave Macmillan.

Duarte-Abadia, B., Boelens, R. and Roa-Avendano, T. (2015). Hydropower, encroachment and the re-patterning of hydrosocial territory: The case of hidrosogamosa in Colombia. *Human Organization* 74(3), pp. 243–254.

Elden, S. (2010). Land, terrain, territory. *Progress in Human Geography* 34(6), pp. 799–817.

Feitelson, E. and Rosenthal, G. (2012). Desalination, space and power: The ramifications of Israel's changing water geography. *Geoforum* 43(2), pp. 272–284.

Forman, G. (2011). Israeli Supreme Court doctrine and the battle over Arab land in Galilee: A vertical reassessment. *Journal of Palestine Studies* 40(4), pp. 24–44.

Galnoor, I. (1978). Water policymaking in Israel. *Policy Analysis* 4(3), pp. 339–367.

Harris, L. M. and Alatout, S. (2010). Negotiating hydro-scales, forging states: Comparison of the upper Tigris/Euphrates and Jordan River basins. *Political Geography* 29(3), pp. 148–156.

Hoogesteger, J., Boelens, R. and Baud, M. (2016). Territorial pluralism: Water users' multi-scalar struggles against state ordering in Ecuador's highlands. *Water International* 41(1), pp. 91–106.

Hoogesteger, J. and Verzijl, A. (2015). Grassroots scalar politics: Insights from peasant water struggles in the Ecuadorian and Peruvian Andes. *Geoforum* 62(5), pp. 13–23.

Ibrahim, A. (2017). *Settler demolition and construction in the Golan*, Jadaliyya. https://tinyurl.com/ydhnbkbw, last accessed May–July 2017 in Arabic.

JNF (2017). *Water for Israel*. www.kkl-jnf.org/water-for-israel/, last accessed March 2017.

Keary, K. (2013). *Water is life*. Al Marsad Publication. http://golan-marsad.org/wp-content/uploads/Water_is_Life_2013.pdf, last accessed March 2017.

Kolers, A. (2009). *Land, conflict, and justice: A political theory of territory*. Cambridge: Cambridge University Press.

Lefebvre, H. (1991). *The production of space*. Oxford: Blackwell.

Lipchin, C. (2007). Water, agriculture and Zionism: Exploring the interface between policy and ideology. In Lipchin. C., Pallant, E., Saranga, D., Amster, A. (eds). *Integrated water resources management and security in the Middle East* (pp. 251–267). Dordrecht: Springer.

Lloyd, D. (2012). Settler colonialism and the state of exception: The example of Palestine/Israel. *Settler Colonial Studies* 2(1), pp. 59–80.

Mara'i, T. and Halabi, U. R. (1992). Life under occupation in the Golan Heights. *Journal of Palestine Studies* 22, pp. 78–93.

Meehan, K. M. (2014). Tool-power: Water infrastructure as wellsprings of state power. *Geoforum* 57(1), pp. 215–224.

Meiton, F. (2015). The radiance of the Jewish national home: Technocapitalism, electrification, and the making of modern Palestine. *Comparative Studies in Society and History* 57(4), pp. 975–1006.

Meiton, F. (2016). Electrifying Jaffa: Boundary-work and the origins of the Arab-Israeli Conflict. *Past & Present* 231(1), pp. 201–236.

Murphy, R. and Gannon, D. (2008). Changing the landscape: Israel's gross violations of international law in the occupied Syrian Golan. *Yearbook of International Humanitarian Law* 11, pp. 139–174.

Ó Cuinn, G. (2011). Land and space in the Golan Heights: A human rights perspective. In Egoz, S. and Pungetti, G. (eds), *The right to landscape: Contesting landscape and human rights* (pp. 85–98). Farnham: Ashgate.

Perreault, T. (2003). 'A people with our own identity': Towards a culural politics of development in Ecuadorian Amazonia. *Environment and Planning D: Society and Space* 21, pp. 583–606.

Ra'ad, B. L. (2010). *Hidden histories: Palestine and the eastern Mediterranean*. London: Pluto Press.

Ram, M. (2015). Colonial conquests and the politics of normalization: The case of the Golan Heights and Northern Cyprus. *Political Geography* 47, pp. 21–32.

Rodgers, D. and O'Neill, B. (2012). Infrastructural violence: Introduction to the special issue. *Ethnography* 13(4), pp. 401–412.

Salamanca, O. J., Qato, M., Rabie, K. and Samour, S. (2012). Past is present: Settler colonialism in Palestine. *Settler Colonial Studies* 2(1), pp. 1–8.

Scott, J. C. (1987). *Weapons of the weak: Everyday forms of peasant resistance*. New Haven: Yale University Press.

Smith, B. J. (1993). *The roots of separatism in Palestine: British economic policy, 1920–1929*. Syracuse: Syracuse University Press.

Star, S. L. (1999). The ethnography of infrastructure. *American Behavioral Scientist* 43(3), pp. 377–391.

Swyngedouw, E. (1999). Modernity and hybridity: Nature, regeneracionismo, and the production of the Spanish waterscape, 1890–1930. *Annals of the Association of American Geographers* 89(3), pp. 443–465.

Swyngedouw, E. (2004). Globalisation or 'glocalisation'? Networks, territories and rescaling. *Cambridge Review of International Affairs* 17(1), pp. 25–48.

Swyngedouw, E. (2009). The political economy and political ecology of the hydro-social cycle. *Journal of Contemporary Water Research and Education* 142, pp. 56–60.

Veracini, L. (2006). *Israel and settler society*. London: Pluto Press.

Veracini, L. (2010). *Settler colonialism: A theoretical overview*. New York: Palgrave Macmillan.

Veracini, L. (2013). The other shift: Colonialism, Israel and the occupation. *Journal of Palestine Studies* 42(2), pp. 26–42.

Weizman, E. (2007). *Hollow land: Israel's architecture of occupation*. London: Verso.

Wessels, J. I. (2015). Challenging hydro-hegemony: Hydro-politics and local resistance in the Golan Heights and the Palestinian territories. *International Journal of Environmental Studies* 72, pp. 601–623.

Wittfogel, K. (1957). *Oriental despotism: A comparative study of total power*. New Haven: Yale University Press.

Wolfe, P. (2006). Settler colonialism and the elimination of the native. *Journal of Genocide Research* 8(4), pp. 387–409.

Yiftachel, O. (2006). *Ethnocracy: Land and identity politics in Israel/Palestine*. Philadelphia, PA: University of Pennsylvania Press.

Zakim, E. (2006). *To build and be built: Landscape, literature, and the construction of Zionist identity*. Philadelphia, PA: University of Pennsylvania Press.

Zeitoun, M. (2008). *Power and water in the Middle East: The hidden politics of the Palestinian-Israeli water conflict*. London: IB Tauris.

Zeitoun, M., Eid-Sabbagh, K., Dajani, M., Talhami, M. (2012). *Hydropolitical baseline of the upper Jordan River. Beirut*: Association of the Friends of Ibrahim Abd El Al.

Zeitoun, M. and Warner, J. (2006). Hydro-hegemony – A framework for analysis of transboundary water conflicts. *Water Policy* 8, pp. 435–460.

Zwarteveen, M. Z. and Boelens, R. (2014). Defining, researching and struggling for water justice: Some conceptual building blocks for research and action. *Water International* 39(2), pp. 143–158.

10 Development initiatives and transboundary water politics in the Talas waterscape (Kyrgyzstan–Kazakhstan)

Towards the Conflicting Borderlands Hydrosocial Cycle[1]

Andrea Zinzani

Introduction

Over the last two decades in the scholarship on the political geography of water and water politics considerable attention has been paid to interactions between water resources management and the state, and in particular on processes of nation-building and state legitimisation through the development of hydraulic projects, infrastructures and the formalisation of water policies (Menga, 2017; Harris and Alatout, 2010; Swyngedouw, 2004; Scott, 1998). This research advanced the understanding of the key role of water resources management for the state by investigating the complex processes of state consolidation, identity formation, scale reconfiguration and national development at the domestic level, and of hydro-hegemony and power relations internationally (Zinzani and Menga, 2017; Warner and Zeitoun, 2008; Sneddon and Fox, 2006). In parallel, with the attempt to reflect on state-society water relations and on multi-scalar approaches, scholars in the political ecology of water have advanced the theorisation of water, power, technology and development policy relations through the design of novel concepts and frameworks (Budds et al., 2014; Linton, 2010; Swyngedouw, 1997). With regard to the Global South and to international development policies, in diverse settings the state has adapted international water initiatives implementation to its own strategies of state legitimisation and bureaucratic reconfiguration (Reis and Mollinga, 2015; Zinzani, 2015; Yalcin and Mollinga, 2007).

Inspired by the scholarship on the political ecology of water and development (Budds and Sultana, 2013), this chapter aims to critically reflect on water, state and development interactions while exploring how international development policies rework hydrosocial relations in transboundary waterscapes. Over the last two decades, international development organisations – the World Bank (WB), the Asian Development Bank (ADB), the United Nations Economic Commission for Europe (UNECE) among others – have supported many governments of transboundary riparian states in the formalisation of interstate agreements based on the United Nations Convention on Non-Navigational Use of International Watercourses (1997) and on the UNECE Water Convention (1992–2003). This support aimed to strengthen interstate water cooperation and reduce potential tensions regarding water demands and allocation. However, beyond formal interstate cooperation, in different contexts states took advantage of the formalisation of water agreements to reconfigure transboundary power relations and to strengthen their bargaining power with riparian states (Zinzani and Menga, 2017; Menga, 2016; Mirumachi, 2015).

By advancing the notion of the Conflicting Borderlands Hydrosocial Cycle, this chapter explores how the establishment of the Chu-Talas Commission, together with riparian states politics, reconfigured the hydrosocial cycle at the borderlands level in the Talas transboundary waterscape, shared by Kyrgyzstan and Kazakhstan. The Chu-Talas Commission, established in 2006, was supported by the UNECE, the United Nations Economic and Social Commission for Asia and the Pacific (UNESCAP) and the Organization for Security and Cooperation in Europe (OSCE). Through the adoption of the hydrosocial cycle and a reflection on the framework in a borderlands setting, this paper analyses the role of state water institutions, their political discourses, and practices of transboundary water infrastructure management. The hydrosocial cycle was conceptualised and developed by Swyngedouw (2009,2004) and Linton (2010) among others, merging the physical and social nature of water and reflecting on interactions between the water flow, power, technology and political transformations. While over the last years the hydrosocial cycle has been reconceptualised and debated by several authors and applied to diverse contexts (Boelens, 2014; Budds et al., 2014; Linton and Budds, 2014; Schmidt, 2014;Budds and Sultana, 2013 among others), its adoption and analysis in transboundary waterscapes remain unexplored. Indeed, there is a lack of research on how the hydrosocial cycle in a transboundary setting might be reconfigured, on the one hand by international development initiatives, while on the other hand by the socio-political logics and politics of riparian states as well as by borderlands practices.

Through the lens of the Conflicting Borderlands Hydrosocial Cycle, this chapter aims at conceptualising a borderlands setting where the hydrosocial cycle both embodies, while being fragmented by, conflicting and divergent water politics in terms of logics, visions and infrastructural property regimes. This specific context emerges in contrast with development organisations' aims of water politics homogenisation and good governance support. Empirical data was collected through a qualitative ethnographic approach: semi-structured interviews, open discussions and informal talks were conducted with members of the Chu-Talas Commission; state, province and district institutions; associations and water users in six villages – Grodikovo, Kyzilkainar and Besagash in Kazakhstan (Dzhambul District, Dzhambul Province) and Uckurgan, Maiska and Talas Aul in Kyrgyzstan (Manas District, Talas Province) – of the Kazakh and Kyrgyz borderlands during August and September 2015. In addition, canals

Development initiatives and transboundary water politics 149

and fields surveys were conducted with members of water institutions, water users and farmers.

This introduction is followed by the next section, which situates the paper in the broader scholarship on the political ecology of water and development, while discussing water, power and development interactions in the context of hydrosocial transformations. Moreover, it presents the hydrosocial cycle and related debates, highlighting the need to analyse the framework in transboundary settings and in borderlands. The second section provides an overview on the Talas transboundary waterscape and explores the process leading to the establishment of the Chu-Talas Commission. Section three presents empirical data and is divided into three sub-parts: the role of state institutions, political discourse and transboundary infrastructure management. The fourth section discusses empirical evidence and interactions between development initiatives, state politics and hydrosocial transformations, while the last one advances the notion of the Conflicting Borderlands Hydrosocial Cycle.

Water, power and development: the hydrosocial cycle and borderlands

Over the last two decades, in parallel with the scholarship on the political geography of water and water politics, a growing body of research has contributed to the political ecology of water (Loftus, 2009; Kaika, 2006; Bakker, 2003; Swyngedouw, 2009, 2004, 1997; Bryant, 1998). Research departed from the assumption that water issues cannot be understood in isolation from the socio-political and economic contexts within which they are produced. As underlined by Swyngedouw (1997) and Budds (2004) among others, this approach enables a deep analysis of power structures and politics that characterise processes of environmental change and socio-political implications of water resource management reconfiguration. Warner and Zeitoun (2008) and Budds and Hinojosa (2012) link this approach with the context of international development. They claim that the political dimension of environmental and climate change has often been ignored by international development actors and experts, who often naturalised water development policies and depoliticised the complex nature of these processes. In relation to the depoliticisation of development policies, the state often repoliticised these measures through their adjustment to national settings and objectives (Reis and Mollinga, 2015; Zinzani, 2015; Veldwisch and Mollinga, 2013).

The reconceptualisation of water as socio-nature enables us to support the argument that social and power relations are not external to water resources but they are embedded (Budds, 2009; Loftus, 2009; Swyngedouw, 2004). Swyngedouw (2004), transcending the modernist idea of nature-society as separate entities, envisions water circulation as a combined physical and social process and as hybridised socio-natural flows that merge together nature and society. This approach advances the understanding of how the state might shape water flows and allocation within waterscapes through institutions, law, politics and hydraulic infrastructural development. Nevertheless, by considering the wave of increasing consolidation of the neoliberal political and economic order in diverse regions at the global level, in particular in Western countries, water politics have been ideologically influenced by the uncontested role of the market and related consensus which promoted state withdrawal, deregulation and commodification of water resources (Budds, 2009; Molle et al., 2009b; Molle, 2008; Cornwall and Brock, 2005; Swyngedouw, 2004; Ferguson, 1990).

Since the 1990s international development organisations, such as the WB and United Nations agencies among other donors, have advocated this approach through the design of projects together with the support of sustainable development and good governance in the Global South, paying particular attention to regions influenced by post-socialist regime changes (Zinzani, 2017,2015; Dukhovny and De Schutter, 2010; Sehring, 2009; Bichsel, 2009). In transboundary waterscapes, in the framework of the UNECE Water Convention (1992–2003) and the United Nations Convention on Non-Navigational Uses of International Watercourses (1997), since the mid-1990s diverse development initiatives have boosted the formalisation of inter-state commissions, national river basin authorities and councils to strengthen cooperation between riparian states. However, international development organisations have often not taken into consideration specific interstate water relations, power asymmetries and national state water politics. Therefore, it is important to critically analyse how international development interventions have shaped and transformed hydrosocial relations, the role of the state and related multi-scalar power reconfigurations.

The adoption of the hydrosocial cycle to transboundary waterscapes and specifically to borderlands, mostly unexplored in political ecology debates, enables us to critically reflect on water, power and development interactions through the analysis of the intersections of state and local politics and practices with those supported by international development organisations. The hydrosocial cycle, conceptualised and debated by Swyngedouw (2009,2004), Budds (2009), Linton (2010) and Linton and Budds (2014) among others furthers and strengthens the theorisation of water, power and technology interfaces and the understanding of how socio-natural transformations of these interfaces rework hydrosocial relations. Swyngedouw (2009,2004) initiated the reflection on the relevancy of linking the transformation of and in the hydrological cycle at the global, regional and local levels on the one hand, and the relations of political, economic, social and cultural power on the other hand. Swyngedouw (2004) defines the hydrosocial cycle as a hybridised socio-natural flow that fuses together nature and society in inseparable ways. Linton (2010) adds that the utilisation of the concept as an analytical framework enables us to interrogate the socio-ecological nature of water, the progressive politics of hydrosocial changes and of dialectical relations between water and society. Reflecting on the hydrosocial cycle's ontological and epistemic nature, according to Schmidt (2014) and Linton and Budds (2014) the cycle has to be considered as a complex geographical and historical process, meaning that the assemblage that produces a particular kind of water and a particular socio-political configuration is always changing. Considering actors with the capacity to rework hydrosocial relations, recent studies conducted in diverse contexts by Bakker (2003), Loftus (2009) and Molle (2008) show that the hydrosocial cycle can be reworked by national or local political and economic reconfigurations, rescaling processes, environmental changes and external international development projects which might rework hydrosocial relations, leading to diverse unexplored contexts.

However, while the hydrosocial cycle was debated and reconceptualised from diverse angles and perspectives, little research paid attention to its adoption and analysis in transboundary waterscapes, in particular where international development initiatives were implemented and specifically at the borderlands level. As stated by Baud and Van Schendel (1997), borderlands can be conceptualised as broad scenes of intense interactions in which people from both sides work out everyday accommodations based on face-to-face relationships. Martinez (1994) advances the notion of the borderlands milieu, stating that ideas, history, traditions and institutions influence the way borderlands communities interact. Newman (2003), emphasising the role of the state, highlights that borderlands, and their development, are determined simultaneously by the political context of the two states and by the social, economic and political interactions between them. Focusing on transboundary water management in Central Asia, Lam (2008) points out that although at the national level non-cooperation is dominant, on the local level interactions between riparian states may prevail, while Wegerich et al. (2012) suggest that transboundary interactions at the borderlands level may occur informally without the acknowledgment of the riparian states' national governments. While on the one hand this literature contributes to conceptualising and understanding borderlands and their relations with the state, on the other hand it has denied the role of international development initiatives in shaping borderlands and their socio-political processes. Therefore, a reflection on the hydrosocial cycle at this level is relevant.

Hydrosocial transformations in the Talas waterscape: the establishment of the Chu–Talas Commission

In Central Asia waterscapes have been developed since ancient times by different empires and political orders and transformed until present times.[2] In the twentieth century the material reification and transformation of these waterscapes have mostly occurred during the Soviet hydraulic mission from the 1950s until the end of the 1980s (Bichsel, 2009; Molle et al., 2009a). Within the scope of these hydraulic development projects the Talas waterscape has been configured since the 1970s. The Talas River originates from the central Kyrgyz Tian Shan Mountains and vanishes, due to its diversion and heavy use in irrigation, by the time it reaches the Moinkum Steppe in Kazakhstan. The reification of the waterscape, in particular its central and downstream parts, has occurred during

152   *Andrea Zinzani*

the early 1970s through the construction of the Kirov Reservoir, located in Kyrgyzstan approximately 15 kilometres upstream of the Kazakh border. Supported by Soviet water authorities, the reservoir was built to regulate the flow of the Talas River to supply Kyrgyz and Kazakh central and downstream parts of the waterscape (Dukhovny and De Schutter, 2010; Wegerich, 2008). Since the mid-1970s other canals have been developed to extend the waterscape and support the establishment of state and collective farms and the reclamation of new irrigated areas.

In 1983, in the framework of Soviet inter-republican water agreements, a Regulation Protocol was signed between the Kyrgyz and Kazakh SSRs in order to formalise an equal division of the Talas water flow, 50% to each republic (Wegerich, 2008). After the collapse of the Soviet Union and the development of different agricultural and economic plans between the two independent states, diverse water demand issues started to emerge. In order to strengthen relations and cooperation between the two states, an agreement "On the use and management of water facilities in the Chu and Talas river basins" was signed in 2000. Parties decided to keep the water division amount accorded during the Soviet Union and water infrastructure operation and maintenance (hereinafter O&M) of intergovernmental status (Kirov Reservoir), to be shared by Kyrgyzstan and Kazakhstan. This agreement came into force in 2002. However, despite its formalisation, since 2002 international development organisations decided to become actively involved in order to strengthen transboundary water cooperation between Kyrgyzstan and Kazakhstan. UNECE, UNESCAP and OSCE promoted, in partnership with the Kyrgyz and Kazakh water authorities, the project "Support for the creation of a commission on the Chu and Talas Rivers between Kazakhstan and Kyrgyzstan", aimed at establishing an interstate basin commission and defining its procedures and costs for exploitation and maintenance of water management infrastructures (Libert, 2014; Libert and Lipponen, 2012). Although water agreement and O&M of water infrastructures were already in place, development organisations saw the opportunity to actively intervene in the interstate political and economic context (UNECE, 2011a).

The project was successfully implemented in 2006, as pointed out by Libert and Lipponen (2012), and the Chu-Talas Commission was established. According to the UNECE (2011b,2011a), the main activities of the Chu-Talas Commission focus on the approval of water resource allocation in the Chu and Talas river basins, the determination of measures to maintain water facilities of interstate use and provide for their capital repair and on the approval of a financing plan for the above measures. Since 2008 UNECE and OSCE have renewed their involvement through the design of a new project, "Developing cooperation on the Chu and Talas river basins", developed in parallel with the ADB and the Swiss Development Cooperation (SDC), aimed at including other water infrastructures under the control of the Commission – the Kozh canal among others – and to introduce Integrated Water Resource Management

Figure 10.1 GIS elaboration of a satellite image representing the upstream part of the Talas waterscape and borderlands, focus of ethnographic research (source: Google Earth TM).

(IWRM) principles into the framework (Libert and Lipponen, 2012; UNECE, 2011a). In order to develop this initiative, international organisations pushed for the establishment of a transboundary basin council with the aim to actively involve, besides state actors, NGOs, communities and water users in decision-making processes through the adoption of good governance principles (ADB, 2013; UNECE, 2011b).

The head of the Kazakh Chu-Talas Basin Water Organisation and the Secretariat of the Chu-Talas Commission, when asked about the transboundary basin council, argued the institutional and political issues related to its establishment and stressed the priority of strengthening joint O&M of water facilities rather than the implementation of IWRM principles. This position was reflected by the head of the Kyrgyz Talas River Basin Authority, who further expressed doubts and a weak sense of conviction regarding the initiative. However, both Kazakh and Kyrgyz water authorities stated that the support of donors was extremely important in establishing the Chu-Talas Commission, to locate a shared water governance vision and to develop strong cooperative ties between Kyrgyz and Kazakh communities. Questioned on the impact of the Commission at the borderlands level, the Secretariat and the members of the Kyrgyz and Kazakh basin

154 *Andrea Zinzani*

water organisations argued that the institutional reconfiguration led to benefits for farmers and water users over the last years, in terms of water allocation, and strengthened collaboration among them. This context seems to demonstrate how the two riparian states were collaborative with international organisations initiatives and their vision.

The hydrosocial cycle in transformation: the Talas borderlands

Water institutions and their roles

Over the last two decades the Kazakh water institutional structure at the sub-national level has been affected by significant changes. According to the Water Code, issued in 2003, the basin and provincial water sectors are managed by two institutions, river basin agencies, branches of the National Committee of Water Resources (Ministry of Agriculture) and based on the river basin principle, and republican state enterprises (*Kazvodkhoz*), patterned by administrative principles (provinces). At the district level water allocation is managed by district water departments (*Kommunalnivodkhoz*) and water users associations (WUAs), established with the law on WUAs in 2004. After a wave of liberalisation promoted by the government since the 2000s, over the last years the institutional structure has been radically changed, influenced by both state and local socio-political and economic dynamics (Zinzani, 2015).

In order to shed light on the impact of the establishment of the Chu-Talas Commission at the borderlands level, water users were asked on related potential benefits, in terms of water amount, use and management. However, farmers and water users interviewed were not aware of the existence of the Commission, even 10 years later following its establishment. The head of the Grodikovo municipality argued that he had heard about the establishment of the Commission only in 2007. Moreover, he further added that over the following years no changes at local level had occurred. Meanwhile, a farmer stated that these institutional and political processes exclusively involved members of national and basin water authorities, excluding members of local administrations and other water users.

However, since 2012 state water politics have significantly impacted the Talas district institutional structure and its hydrosocial relations. In 2012 the Taraz *Kazvodkhoz* strengthened its institutional and political role when its control and financial support shifted from the province to the state. In addition, this state authority was appointed to be the only institution to manage water allocation in the province. When asked about this process, members of the Taraz *Kazvodkhoz* stated that since 2014 the Dzhambul province, supported by state water authorities, has required the dismantlement of the *Kommunalnivodkhoz* and the recentralisation of district water allocation and hydraulic infrastructures under the supervision of the *Kazvodkhoz*. In fact, few years ago the Dzhambul district *Kommunalnivodkhoz* was in the process of

Development initiatives and transboundary water politics   155

dismantlement, and is expected to be replaced by a district branch of the Taraz *Kazvodkhoz*. Both members of the Taraz *Kazvodkhoz* and the Talas River Basin Agency claimed that this process was motivated by economic and technical concerns: on the one hand the district government lacked resources to support the *Kommunalnivodkhoz*, while on the other hand this issue implied lack of infrastructure maintenance and control, a decrease of the whole irrigated land and problems of water allocation. With regard to the property regime of secondary canals, 21 of them shifted from district to state control, the Kozh canal among them. Differently, the Tyute canal, privatised in 2004 and currently managed by a private enterprise, was in the process of being renationalised, while the Imankul canal, property of Kyzilkainar municipality, will be considered for nationalisation in the forthcoming years. Therefore, if we consider water allocation procedures, after the dismantlement of the Dzhambul *Kommunalnivodkhoz*, state water authorities will be responsible for this service. This specific reorganisation occurred since any WUA has been operating in the area despite WUAs national formalisation in 2004. As stressed by an employee of the Grodikovo municipality and a farmer of Besagash, two WUAs operated in the district until three years ago but they filed for bankruptcy while two others that were planned were not created due to administrative and bureaucratic issues (Zinzani, 2015). Farmers in Grodikovo and Besagash stated that the canal property regime shift and state water management had already improved O&M, that water fees were lower and that soon further state funds will be needed to cover infrastructural repairs.

Also in Kyrgyzstan the water institutional context has evolved over the last two decades, in particular at the district level. Water resources at provincial level are managed by river basin authorities, branches of the Department of Water Economy and Melioration (Ministry of Agriculture). While on the one hand these authorities were responsible of water amount regulations and O&M of trans-districts canals, on the other hand, at the district level, district water departments (*Rayvodkhoz*) were in charge of district main canals O&M and water allocation to WUAs, supported since 2000 by the WB. In parallel, the WB promoted the formalisation of a National Department for WUAs with supportive branches at the provincial and district levels. Furthermore, since 2008 in some areas federations of WUAs have been established by water users with the aim of managing a portion of the main canals leased by district water departments for a period of four years. When asked about the establishment of the Chu-Talas Commission, the head of the Manas District Water Department, responsible for O&M of the district's main canals, stated that they heard about the process and the involvement of the Ministry and some members of the Talas Basin Authority and added that his department was not involved. It seems that the process primarily involved only the Ministry and state authorities. With regard to the impact of the establishment of the Commission at the borderlands level, the majority of farmers and water users interviewed in Uckurgan, Maiska and Talas Aul were not even aware of its existence. Only two of them, one former head of the Talas Aul municipality

156 *Andrea Zinzani*

and the other a former worker at the Kirov Reservoir, were informed of its establishment almost a decade ago. They added that this process did not lead to significant changes concerning water amount availability, or on water allocation procedures.

However, two institutional reconfigurations led to rather significant hydrosocial transformations in Kyrgyzstan. The first was a project of the WB, attempted on the one hand to support WUAs while on the other hand to establish a national authority to support their development. The second was based on the claim that heads of WUAs were called on to manage the main canals, replacing the Manas District Water Department, thereby establishing federations of WUAs. Since 2002, 11 WUAs have been established in the district due to the initial support of the WB, and afterwards due to the creation of the Manas District Department of WUAs support. Its director, a former member of the Manas District Water Department, stated that the creation of WUAs gave farmers the opportunity to maintain their canals and to organise water allocation by themselves. A member of the Dikhan WUA's board argued that the WUA initiative enables farmers' participation in water decision-making procedures and their self-accountability regarding water use, payment and saving procedures. However, despite this institutional context, evidence from informal talks and meetings showed that Dikhan, Maksat-A and Manas-C WUAs are directed by former heads of state and collective farms and former members of municipalities, and that participatory processes are strictly related to specific social dynamics and practices. Furthermore, as stressed by the Maksat-A WUA director and by farmers of the Dikhan WUA, WUAs would not be financially sustainable without the support of municipalities.

Another relevant hydrosocial reconfiguration has occurred since 2008, when four WUAs of the district created the Saiza-Baisu WUAs Federation, headed by the director of the Mamatbaisu WUA. He concurred that the Federation was designed following the request of water users and heads of other WUAs to deal with the lack of canal maintenance and water allocation efficiency of the Manas District Water Department. Two main canals, the Saiza and the Baisu, were leased from the Department for a period of four years. Considering the reconfiguration of roles and power in water control, the design of the Saiza-Baisu Federation reduced the role and the amount of funds – delivered by the Ministry – to the Manas District Water Department. When asked about the potential establishment of other federations in the district, the director of Saiza-Baisu Federation argued that the creation of new federations would be a serious institutional challenge since it would significantly decrease the role of the district water department, which is today the unique state water institution in the district. According to him, it seems that any other WUAs federation would be created in the future in order to keep the role and power of the state and private organisations in balance. In sum, these dynamics depict how local political elites and the power of former state officials were relevant in shaping the processes of WUAs and their relations with district-level state water authorities.

The political discourse

The formalisation process of the Chu-Talas Commission was characterised by discourses of development organisations based on depoliticised narratives and development buzzwords such as good governance, knowledge sharing and capacity building. These discourses and their visions, narrated by development organisations to Kazakh and Kyrgyz states and water authorities, have been then shaped, re-nationalised and spread to provincial and district levels. Kazakh water authorities spread a specific political discourse based on institutional and power transformations analysed in the previous section.

With regard to the political vision of the process, the Kazakh state and its water authorities supported the idea of a reconsideration of water liberalisation that occurred over the last decade and the strategy to recentralise water management. When asked about the institutional recentralisation and the shift in infrastructural property regime, members of the Taraz *Kazvodkhoz* argued that canals and water allocation procedures are strategic objectives that should be managed exclusively by the state, underlining that the decentralisation of secondary canal control failed, both financially as well as technically. In addition, they argued that cases of infrastructure privatisation, as for instance the Tyute canal, led to excessive commodification of water resources, private capital accumulation and inequalities among users. The director of the Chu-Talas River Basin Agency argued that nowadays the Kazakh state possesses the financial capacity to re-nationalise infrastructures and that the dismantlement of authorities operating between the state and water users will decrease water fees and increase subsidies to farmers. He added that Kazakh water decentralisation failed to provide technical and social improvements, while farmers in Grodikovo stated that the decentralisation did not lead to benefits and that the WUAs experience was not successful. Water users interviewed in Besagash and a former technician of *Kommunalnivodkhoz* agreed that if state authorities manage infrastructures, their control of water allocation would be more efficient. This context clearly shows how the discourse supported by state water authorities has permeated water users' ideas and visions, still negatively influenced by the outcomes of the recent liberalisation wave.

In contrast, in Kyrgyzstan state water authorities have mostly shared and pushed narratives of development organisations and in particular those of the WB. In fact, since 2002, when the WB initiated the establishment of WUAs, the discourse of Kyrgyz water authorities has spread decentralisation, liberalisation and participatory approaches. The director of the Manas District Water Department remarked that agreements between the government and the WB demonstrated their organisational and technical commitment to establish e district departments of WUAs support. When asked about current policies and visions for the near future, he and other two members argued that, with the exception of main infrastructures, the other canals belong to the people and not to the state, and that water users through WUAs should be held responsible

158 *Andrea Zinzani*

for the maintenance of their infrastructures. Concerning the creation of the Saiza-Baisu WUAs Federation, members of the Manas District Water Authority stated that the request for its establishment was accepted and that the process was inspired by their belief in decentralisation and by the idea that state and private institutions could operate together without conflicts of interest. Moreover, water users argued that decentralised management enables their active involvement in decision-making procedures.

However, informal talks with key informants showed that WUAs mechanisms were deeply influenced by local bureaucracies, by the power of municipalities and in certain cases by members of the Manas District Water Department. Moreover, it emerged that the financial support of municipalities was essential for WUAs and that without their support it will be challenging for them not to file for bankruptcy. Therefore, despite a state political discourse inspired by decentralisation and liberalisation, Kyrgyz authorities and bureaucracies still played a powerful role in influencing borderlands water management and politics.

Borderlands management and practices of transboundary infrastructures

Borderlands of the Talas waterscape are crossed by canals connected both to the Kirov Reservoir and to the river. The nature of the waterscape, combined with heterogeneous socio-political and technical dynamics, have led to diverse transboundary issues in terms of water amount sharing, use and management of infrastructures.

Research findings from borderlands showed that O&M of these small infrastructures was not included in the framework of the Commission. This specific physical and institutional context led to the rise of tensions and disputes at the level of borderlands on the one hand, and of formal and informal practices of cooperation between institutions and actors on the other hand.

The Kozh canal represents an interesting case in point for the analysis of infrastructural and socio-power issues within the hydrosocial cycle. The canal is linked to the Talas River in the Kyrgyz territory and measures approximately 10 kilometres; three kilometers are in Kyrgyzstan, while the other seven are in Kazakhstan. Its property regime and management reflects the diverse water political vision of the two states. The Kyrgyz section is controlled by the Maksat-A WUA, while the Kazakh one since 2014 by the Taraz *Kazvodkhoz*. This configuration makes the canal management a complex challenge. A technician of Taraz *Kazvodkhoz* stated that technical, maintenance and water allocation issues often occur. Furthermore, he added that technicians are in contact with the director of Maksat-A WUA and in cases of technical issues in the Kyrgyz section of the canal they are allowed to cross the border in order to negotiate with them and redress particular issues. However, due to the institutional nature of the canal, an international agreement is not possible. When

Development initiatives and transboundary water politics 159

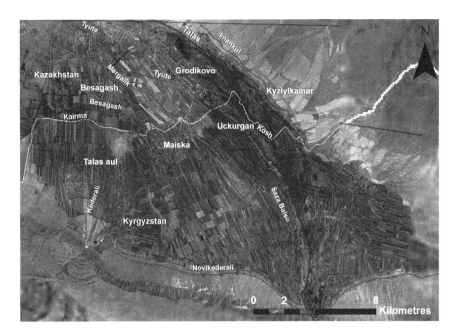

Figure 10.2 GIS elaboration of a satellite image representing the Talas waterscape borderlands, the Talas River and the canals network, and villages where ethnographic research was conducted (source: Google Earth TM).

asked about this context, the director of the Maksat-A WUA affirmed that different issues were at stake. First 1.6 kilometres of the canal were lined, while the following 1.2 kilometres were grounded. These specific infrastructural characteristics often led to flooding issues in the fields of farmers located close to the border, but the WUA's small budget does not enable structural renovations or maintenance in the part close to the border. The director stated that at the beginning of the vegetation season, *Kazvodkhoz* technicians often cooperate with the WUA's members for the canal's maintenance. In addition, he argued that since the end of the Kyrgyz part of the Kozh canal provides irrigation exclusively for land located in Kazakhstan, the Kazakh government should be in charge of this section. However, Kazakh authorities pointed out that it lies in the Kyrgyz territory, and that they already pay for joint infrastructures in the framework of the Commission. In parallel, the Manas District Water Department, despite official requests from the Maksat-A WUA, was not willing to be responsible for the canal and its related issues. The WUA director, who argued that the Kozh canal should be included in the interstate strategic infrastructures in the framework of the Commission, expressed his claims on technical issues both to the Talas River Basin Authority and the Commission, and received

160   *Andrea Zinzani*

no reply. According to reports of the Commission and the UNECE, the canal should have been included in the list of interstate strategic infrastructures since 2011, but evidence from borderlands revealed a different and highly contested context.

Diversely, the Kairma stream, which originates close to the border, flows in Kyrgyzstan parallel to the border for approximately 5 kilometres. According to the head of a farmers cooperative located in Kazakh borderlands, since 2012 the Kairma was informally connected to their lands through transboundary underground pipes. When asked about relations with Kyrgyz authorities, he maintained that he did not have contacts with members of the Dikhan WUA, or with those of the Manas District Water Department. However, informal cooperation practices between the Kazakh cooperative and the Maiska municipality emerged. He further revealed that since some years he had stipulated an informal agreement with members of the municipality through a regular payment for water use. In order to shed light and understand this informal practice, members of the Maiska municipality were asked about these relations with the Kazakh cooperative; they denied having contact and any kind of agreement with Kazakh cooperatives and their farmers.

In addition, claims on water use asymmetries with Kazakhstan interestingly emerged from discussions and informal talks with heads of Kyrgyz WUAs, together with farmers. They argued that Kazakh authorities should pay for water which flows from Kyrgyz springs (Besagash and Kairma) to their territories and for flooding issues along the Kozh canal and in Maiska village. The head of Manas-C WUA argued that Kazakhstan receives water for free and that its authorities immediately complain in case of a decrease in water flow; he added that these issues should be regulated at the governmental level. Moreover, the Deputy of Maiska, the former head of the Dikhan WUA, stated that he contacted the Talas River Basin Authority to put pressure on the Ministry in order to discuss and try to solve these issues in the framework of the Chu-Talas Commission. However the claim remained unanswered.

The hydrosocial cycle between development initiatives, state politics and borderlands practices

The analysis shows that the establishment of the Chu-Talas Commission had very little influence in the Kazakh-Kyrgyz borderlands hydrosocial cycle. District-level water authorities were excluded from the process of establishing the Commission, and were not informed by national- and basin-level state authorities. It emerged that the majority of members of district-level authorities are not aware of the existence of the Chu-Talas Commission, even 10 years after its formation. Since heads and members of the WUAs and farmers were also not aware of its existence, its potential benefits at the borderlands level in terms of water amount and allocation, underlined by Kazakh state authorities, seem doubtful. The strategy of knowledge sharing, good governance, IWRM principles support, pushed by development organisations since

Development initiatives and transboundary water politics 161

2008, has been almost ignored by governments, in particular by the Kazakh state. In parallel, also the establishment of the transboundary basin council was not even considered by both the Commission and the Kazakh and Kyrgyz national water authorities. Its potential establishment would have significantly reconfigured borderlands power relations, making political interactions challenging between the two states, in particular if we consider conflicting water political visions undertaken by the two states over the last few years. Thus, the Kazakh and Kyrgyz state authorities were able to adapt logics and discourses promoted by development organisations to meet their own objectives and political strategies.

Therefore, the borderlands hydrosocial cycle, rather than being reconfigured by the establishment of the Commission, was shaped by national and provincial state political agendas and borderlands informal practices. The analysis of the roles of institutions and their discourses enabled the understanding of these processes. In Kazakhstan, a national initiative led to an institutional transformation process – the dismantlement of the Dzhambul *Kommunalnivodkhoz* and related reconfiguration of the role of the Taraz *Kazvodkhoz* – oriented towards a water state recentralisation and characterised by a significant shift of power aimed at strengthening the centrality of the state. The process was also bolstered by a political discourse based on the mistrust of decentralisation and on the failure of liberalisation, underlying the organisational and technical incapacities of district-level authorities, consequences of privatisation and the bankruptcy of certain WUAs. Furthermore, state water authorities highlighted the strategic role of infrastructures and the imperative that they should be the property of the state. This vision was presented to water users as a regime based on certainty, equality and lower charges, despite the inequalities and loss of power that affected former members of the Dzhambul *Kommunalnivodkhoz*. Water users, having abandoned the idea to create WUAs, an initiative that stood in contrast to the state recentralisation, are today directly related to state authorities in terms of water allocation schedules. With regard to Kyrgyz authorities, this political vision enables the Kazakh state to maintain a more powerful position in terms of transboundary negotiations and cooperation (Zinzani and Menga, 2017).

Diversely, in Kyrgyzstan the borderlands hydrosocial cycle was mostly influenced by the WB political vision that led to the establishment of WUAs and federations of WUAs. Since 2002 the Kyrgyz government has boosted projects and initiatives promoted by the WB, for example by establishing a new state department (Department for WUAs Support) at the national, provincial and district levels in order to formalise the process. Kyrgyz state water authorities, contrary to those of Kazakhstan, supported a narrative centered on the belief in liberalisation and on the idea that water and infrastructures should be managed according to a public-private partnership. From the perspective of water users, the reconfiguration process enabled them to be part of associations and to be involved in decision-making procedures. The creation of federations of WUAs demonstrated the power of water users in

162 *Andrea Zinzani*

terms of decision-making, capacities of influencing state policies and the reduction of the role of the state. However, despite this context, members of municipalities, bureaucrats of district water authorities and former members of collective farms still play a relevant role in influencing decisions and power dynamics of WUAs. With regard to transboundary relations with the Kazakh state, Kyrgyzstan showed weaker negotiation capacities and bargaining power with respect to Kazakhstan, also in relation to the claims made by borderlands water users, which were mostly ignored by both Kyrgyz authorities and the Commission.

Towards the Conflicting Borderlands Hydrosocial Cycle

This contribution aimed at analysing how the establishment of the Chu-Talas Commission, together with riparian state politics, reworked the hydrosocial cycle in the borderlands of the Talas waterscape. The analysis of the role of state institutions, their political discourses and the management of transboundary infrastructures enabled the understanding of the centrality of the state and its authorities, rather than of development organisations, in the hydrosocial cycle reconfiguration. Indeed, the Talas waterscape borderlands hydrosocial cycle was reworked, in terms of its socio-power transformations, by state national and sub-national politics, and partly by borderlands practices rather than by initiatives, policies and narratives of international development organisations that were promoted since 2002. Although riparian states and actors seem to share the same water flow and its use (via irrigation), state politics and the infrastructural property regimes between Kazakhstan and Kyrgyzstan deeply differ and sit in contrast to one another in terms of their political logics, rules, resource management visions and state–private actors power relations. The recent process of the hydrosocial cycle reconfiguration deeply contrasts with the vision of development organisations, which aimed at homogenising and structuring borderlands through the support of neoliberal principles and related narratives. Therefore, rather than a homogeneous interstate context shared by riparian states, the evidence shown enables us to argue that the Talas waterscape borderlands hydrosocial cycle reveals a complex heterogeneous and contrasting nature, characterised by opposing and conflicting political rationalities, processes and discourses, and roles of institutions and communities.

With regard to this contested context, this chapter aims to provide a novel contribution to the understanding of the hydrosocial cycle in transboundary settings. The nature of the dynamics that emerged from the Talas waterscape borderlands enables a preliminary conceptualisation of the Conflicting Borderlands Hydrosocial Cycle. This notion embodies the contradictory roles of riparian state institutions, political visions and discourses, together with their infrastructural property regimes. Moreover, if we consider the multi-dimensional nature of the hydrosocial cycle, the proposed notion enables a reflection on the fragmented nature of the hydrosocial cycle that pervades very different realities,

Development initiatives and transboundary water politics 163

illustrating that borderlands and the border matter. Despite the fact that conflicts and power contestations are frequently embedded in the hydrosocial cycle as highlighted by Swyngedouw (2004) and Budds and Hinojosa (2012) among others, the Conflicting Borderlands Hydrosocial Cycle emphasises the role of the border in physically separating two contradictory visions and practices of water politics. In relation to the border, despite the rise of heterogeneous informal practices on transboundary infrastructures, a borderlands milieu, as conceptualised by Martinez (1994), has not emerged. Indeed, Kazakhstan and Kyrgyzstan seem to cooperate more at the interstate level within the framework of the Chu-Talas Commission rather than at the borderlands level, as the non-establishment of the transboundary basin council, mostly hampered by Kazakh state authorities, demonstrates.

In conclusion, despite the fact that the UNECE and other organisations perceive the Chu-Talas Commission "as a success story and model for cooperation in Central Asia" this contribution enables the understanding of the ambiguities between narratives emanated by international development organisations, the central role of riparian state politics and their related contested processes, thereby advancing the notion of a Conflicting Borderlands Hydrosocial Cycle. More research across diverse socio-spatial contexts would allow for the strengthening and debate of this notion and offer novel insights for its further development.

Notes

1 This chapter is a revised version of an earlier article published in FMSH Working Papers (Zinzani, 2017); part of the empirical data has been used in an article published in *Geoforum* (Zinzani and Menga, 2017).
2 The waterscape is not merely the territory within which water flows but a produced socio-natural entity in which social power is embedded and shaped by the material flow of water as well as by the assemblage of infrastructures, institutions and authorities, formal and informal practices, discourses and narratives (Swyngedouw, 1997; Budds and Hinojosa, 2012).

References

Asian Development Bank. 2013. *Regional Technical Assistance: Improved Management of Water Resources in Central Asia – Under Activity 3: Support of Chu-Talas Commission.* Technical Assistant's Consultant Reports.

Bakker, K. 2003. A Political Ecology of Water Privatization. *Studies in Political Economy.* 70, pp. 35–58.

Baud, M. and Van Schendel, W. 1997. Toward a Comparative History of Borderlands. *Journal of World History.* 8(2), pp. 211–242.

Bichsel, C. 2009. *Conflicts Transformation in Central Asia: Irrigation Disputes in the Fergana Valley.* London and New York: Routledge.

Boelens, R. 2014. Cultural Politics and the Hydrosocial Cycle: Water, Power and Identity in the Andean Highlands. *Geoforum.* 57, pp. 234–247.

164 *Andrea Zinzani*

Bryant, R. 1998. Power, Knowledge and Political Ecology in the Third World. *Progress in Physical Geography.* 22(1), pp. 79–94.

Budds, J. 2004. Power, Nature and Liberalism: the Political Ecology of Water in Chile. *Singapore Journal of Tropical Geography.* 25(3), pp. 122–139.

Budds, J. 2009. Contested H20: Science, Policy and Politics in Water Resources Management in Chile. *Geoforum.* 40, pp. 418–430.

Budds, J. and Hinojosa, L. 2012. Restructuring and Rescaling Water Governance in Mining Contexts: The Coproduction of Waterscapes in Perù. *Water Alternatives.* 5(1), pp. 119–137.

Budds, J., Linton, J. and McDonnell, R. 2014. Editorial – The Hydrosocial Cycle. *Geoforum.* 57, pp. 167–169.

Budds, J. and Sultana, F. 2013. Editorial: Exploring Political Ecologies of Water and Development. *Environment & Planning D.* 31, pp. 275–279.

Cornwall, A. and Brock, K. 2005. What Do Buzzwords do for Development Policy? A Critical Look at 'Participation', 'Empowerment' and 'Poverty Reduction'. *Third World Quarterly.* 26(7), pp. 1053–1060.

Dukhovny, V. and De Schutter, J. 2010. *Water in Central Asia: Past, Present and Future.* London: CRC Press.

Ferguson, J. 1990. *The Anti-Politics Machine: "Development", Depoliticization and Bureaucratic Power in Lesotho.* Cambridge: Cambridge University Press

Harris, L.M. and Alatout, S. 2010. Negotiating Hydro-scales, Forging States: Comparison of the Upper Tigris/Euphrates and Jordan River Basins. *Political Geography.* 29(3), pp. 148–156.

Kaika, M. 2006. Dams as Symbols of Modernization: The Urbanization of Nature Between Geographical Imagination and Materiality. *Annals of the Association of the American Geographers.* 96(2), pp. 276–301.

Lam, S. 2008. *Cooperation in the Ferghana Valley Borderlands: Habitus, Affinity, Networks, Conditions.* Ph.D. thesis, Department of Geography, Kings College of London.

Libert, B. 2014. The UNECE Water Convention and the Development of Transboundary Water Cooperation in the Chu-Talas, Kura, Drin and Dniester River Basins. *Water International.* 40(1), pp. 168–182.

Libert, B. and Lipponen, A. 2012. Challenges and Opportunities for Transboundary Water Cooperation in Central Asia: Findings From UNECE's Regional Assessment and Project Work. *International Journal of Water Resources Development.* 28(3), pp. 565–573.

Linton, J. 2010. *What Is Water? The History of a Modern Abstraction.* Vancouver: University of British Columbia Press.

Linton, J. and Budds, J. 2014. The Hydrosocial Cycle: Defining and Mobilizing a Relational-Dialectal Approach to Water. *Geoforum.* 57, pp. 170–180.

Loftus, A. 2009. Rethinking Political Ecology of Water. *Third World Quarterly.* 30(5), pp. 953–968.

Martinez, O.J. 1994. *Border People: Life and Society in the U.S.-Mexico Borderlands.* Tucson: University of Arizona Press.

Menga, F. 2016. Reconceptualizing Hegemony: The Circle of Hydro-hegemony. *Water Policy.* 18(2), pp. 401–418.

Menga, F. 2017. Hydropolis: Reinterpreting the Polis in Water Politics. *Political Geography.* 60, pp. 100–109.

Mirumachi, N. 2015. *Transboundary Water Politics in the Developing World.* London: Routledge.

Molle, F. 2008. Nirvana Concept, Narratives and Policy Models: Insights From the Water Sector. *Water Alternatives.* 1(1), pp. 131–156.

Molle, F., Foran, T. and Kakonen M. 2009b. *Contested Waterscapes in the Mekong Region.* London: Routledge Earthscan.

Molle, F., Mollinga, P.P. and Wester, P.2009a. Hydraulic Bureaucracies and the Hydraulic Mission: Flows of Water, Flows of Power. *Water Alternatives.* 2(3), pp. 328–349.

Newman, D. 2003. On Borders and Power: A Theoretical Framework. *Journal of Borderlands Studies.* 18(1), 13–25.

Reis, N. and Mollinga, P.P. 2015. Public Policy and the Idea of the Vietnamese State. The Cultural Political Economy of Domestic Water Supply. *Asian Studies Review.* 39(4), pp. 628–648.

Schmidt, J.J. 2014. Historicising the Hydrosocial Cycle. *Water Alternatives.* 7(1), pp. 220–234.

Scott, J. 1998. *Seeing Like a State: How Certain Schemes to Improve the Human Condition Have Failed.* New Haven, CT: Yale University Press.

Sehring, J. 2009. Path Dependencies and Institutional Bricolage in Post-Soviet Water Governance. *Water Alternatives.* 2(1), pp. 61–81.

Sneddon, C. and Fox, C. 2006. Rethinking Transboundary Waters: A Critical Hydropolitics of the Mekong Basin. *Political Geography.* 26(2), pp. 181–202.

Swyngedouw, E. 1997. Exploring Human Geography: A Reader. *Transactions of the Institute of British* Geographers. 22(4), pp. 237–254.

Swyngedouw, E. 2004. *Social Power and the Urbanization of Water.* Oxford: Oxford University Press.

Swyngedouw, E. 2009. The Political Economy and the Political Ecology of the Hydro-social Cycle. *Journal of Contemporary Water Research & Education.* 131, pp. 56–60.

United Nations.1997. Convention on the Law for Non-Navigational Uses of International Watercourses. Document 8 (2), pp. 1–154.

United Nations Economic Commission for Europe.2011a. *Strengthening Water Management and Transboundary Water Cooperation in Central Asia. The Role of UNECE Environmental Conventions.* Working Paper 3, pp. 1–81.

United Nations Economic Commission of Europe.2011b. *Development of Cooperation in the Chu-Talas Rivers* (Chu-Talas II). Project Report, pp. 1–9.

Veldwisch, G. and Mollinga, P.P. 2013. Lost in Transition? The Introduction of Water Users Associations in Uzbekistan. *Water International.* 38(6), pp. 758–773.

Warner, J.F. and Zeitoun, M. 2008. International Relations Theory and Water Do Mix: A Response to Furlong's Troubled Waters, Hydro-hegemony and International Water Relations. *Political Geography.*27(7), pp. 802–810.

Wegerich, K. 2008. Passing Over the Conflict: The Chu Talas Basin Agreement as a Model for Central Asia. In Rahaman, M.M. and Varis, O. Ed. *Central Asian Waters.* Helsinki: Helsinki University of Technology-Water & Development Publications, pp. 117–131.

Wegerich, K., Kazbekov, J., Kabilov, F. and Mukhamedova, F. 2012. Meso-Level Cooperation on Transboundary Tributaries and Infrastructures in the Fergana Valley. *International Journal of Water Resources Development.* 28(3), pp. 525–543.

Yalcin, R. and Mollinga, P. 2007. *Institutional Transformations in Uzbekistan's Agricultural and Water Resources Administration: The Creation of a New Bureaucracy.* ZEF Working Paper Series 22. University of Bonn.

Zinzani, A. 2015. *The Logics of Water Policies in Central Asia: The IWRM Implementation in Uzbekistan and Kazakhstan.* Zurich and Berlin: LIT Verlag – Series Geographie.

Zinzani, A. 2017. Beyond Transboundary Water Cooperation: Rescaling Processes and the Hydrosocial Cycle Reconfiguration in the Talas Waterscape (Kyrgyzstan, Kazakhstan). *FMSH Working Papers Series*.123, pp. 1–24.

Zinzani, A. and Menga, F. 2017. The Circle of Hydro-Hegemony Between Riparian States, Development Policies and Borderlands: Evidence From the Talas Waterscape (Kazakhstan-Kyrgyzstan). *Geoforum*. 85, pp. 112–121.

11 Speculation and seismicity

Reconfiguring the hydropower future in post-earthquake Nepal

Austin Lord

Anticipating a hydropower future

The Chilime Hydropower Company conducted Nepal's first 'local shares offering' in 2010, following a prolonged struggle with locals from the Rasuwa district who were impacted by the construction of the Chilime Hydropower Project.[1] In technical terms, the offering proved a massive success: local participation exceeded expectations, the logistics of enrollment went smoothly, and share prices rose. Kul Man Ghising, former Chilime CEO, later recalled the scale of local demand for shares in evocative detail: "The poor villagers invested more than NRs 15 crore [~USD $1.5 million] in cash . . . Even the poorest of the poor came forward to apply for the IPO [initial public offering] with old crumpled currency notes and even coins" (ShareSansar, 2014). The people of Rasuwa, who have long struggled against the extractive propensities of the Nepalese state and who maintain their own critical perspectives on development projects in the region (Campbell, 2010), were perhaps understandably interested in the prospect of "becoming an agent of *bikas* [development] rather than one of its targets" (Pigg, 1992: 511). By the time the share certificates were distributed, a new model for hydropower development had been created, and more than 30,000 'project-affected people' had been transformed into local investors.[2]

In Nepal, the imagined hydropower future pervades the uncertain present, giving shape to an indefinite economy of anticipation (cf. Adams et al., 2009; Cross, 2015) animated by recursive patterns of hype and hope (cf. Gyawali, 2006).[3] For decades, the promise that hydropower development will someday prove to be Nepal's "passport out of poverty" (Dixit and Gyawali, 2010: 106) has dominated public discourse, nurturing dreams of becoming a 'hydropower nation' (Lord, 2014).[4] These conditions of extended indeterminacy have fostered a palpable sense of expectation among Nepalis living in the vicinity of proposed hydropower projects, many of whom desire the changes that hydropower development might bring (Rest, 2012) and seek recognition of their rights as 'project-affected people' (Lord, 2016). When the Chilime Hydropower Company developed 'the shareholder model' in response to local demands to be recognized as legitimate stakeholders in the promised hydropower future, they created a way to translate these diverse aspirations into capital. In the years since, more than

168 *Austin Lord*

a dozen hydropower companies have conducted similar offerings, and almost one million Nepalis have purchased shares in the imagined hydropower future.[5]

Anticipation is a particular kind of affective state generated by "the palpable effect of the speculative future on the present" (Adams et al., 2009: 247). 'Economies of anticipation' emerge from the entanglement of diverse anticipatory practices that produce certain spatial and temporal frames, creating a shared forum in which "expected or promised futures conflict or converge" (Cross, 2015: 426). In this sense, "anticipation is not just betting on the future; it is a moral economy in which the future sets the conditions of possibility for action in the present, in which the future is inhabited in the present" (Adams et al., 2009: 249) In this paper, I explore the affective and temporal politics that animate dreams of becoming a 'hydropower nation' in Nepal, focusing on the ways that the state and the hydropower sector have used *capitalist technologies of imagination* (Bear, 2015a) to coordinate speculative visions of an imagined hydropower future.[6] Drawing on the promissory logics of finance, these technologies are used to promote popular speculation and to create a particular kind of 'infrastructural public' (Collier et al., 2016): the citizen-shareholder of the imagined 'hydropower nation'.

Writing about the experiences of people waiting for a 'future oil economy' to arrive in São Tomé and Príncipe, Weszkalnys (2016) has recently introduced the concept of '*resource affect*', stating that affect "is less an externality of resource extraction to be explained away than constitutive of its social-material fabric" (p. 141).[7] Arguing that 'resource affect' is central in political struggles over resource extraction, she highlights a new set of corporate policies and state practices which focus "no longer simply on macroeconomic issues and elite politics but on the purported hopes, desires, and aspirations of citizens in producer states" (ibid). My own analysis foregrounds a similar pattern in Nepal, wherein state and corporate actors seek to cultivate and privatize particular forms of 'resource affect', so as to promote the "productive involvement of the citizens of resource-rich countries in the creation of resource wealth" (p. 140). As anticipatory visions of the 'hydropower nation' wax and wane, I consider the ways that state and corporate actors seek to modulate and capitalize on Nepali patterns of resource affect in unstable times.[8]

More concretely, my analysis concentrates on the 7.8 magnitude earthquake that struck Nepal on April 25, 2015 and the ways this event did or did not alter the shape of its imagined hydropower future.[9] Disasters, crises, and breakdowns often highlight infrastructural politics, prompting debate about the risks posed by both the presence and absence of infrastructures (cf. Carse, 2017; Howe et al., 2016; Larkin, 2013). Debates about seismicity, environmental hazards, and dams are ongoing across the Himalayan region (Gergan, 2016; Drew, 2017; Huber et al., 2017). Similarly, the 2015 earthquake highlighted the precarity of present infrastructures and stirred debate about 'the damage done and the dams to come' in Nepal (Rest et al., 2015).

In the wake of the 2015 earthquake, Critics and public officials openly questioned the centrality of hydropower within Nepal's imagined energy future, highlighting the need to balance desires for energy security and hydropower revenues with the inherent risks of building dams in a region where earthquakes and geohazards are a common occurrence (Hilton, 2015; Thapa and Shrestha, 2015). And yet, though policymakers and representatives of the hydropower sector initially expressed interest in disaster risk reduction, they also remained focused on ensuring that Nepal's hydropower frontier remained open for business. The logics of capital deployed to finance the building of a 'hydropower nation' demanded continuity. Slowly, construction resumed, and a new wave of hydropower IPOs was met with overwhelming demand (ShareSansar, 2016, 2015). Popular 'dreams of hydropower dollars' (Bhushal, 2016) began to circulate anew.

As the aftermath continued to unfold, familiar economic urgencies and geopolitical concerns emerged. In late 2015, an 'unofficial blockade' along the Nepal-India border caused a fuel crisis, eliciting anxieties about energy insecurity and national 'energy sovereignty' (Pattison, 2015; Sangraula, 2015). Using affective rhetoric to highlight chronic insecurities and threats, the state and its corporate allies invoked a moral and pre-emptive narrative of security that would subordinate the unsteady past and "render the future geographies of infrastructure actionable" (Anderson, 2010: 785). In early 2016, the government of Nepal declared an 'energy emergency' and propagated an official crisis management plan organized around the slogan: 'Nepalko Paani, Janatako Lagaani' or 'Nepal's Water, the People's Investment' (Government of Nepal, 2016). Combining affective nationalist rhetoric with an assertion of the productive possibilities of an imagined people-public-private partnership, this slogan epitomizes the core concerns that animate this recalibrated anticipatory regime.

Highlighting the pre-emptive quality of anticipation, Adams et al. (2009) suggested that "anticipatory regimes offer a future that may or may not arrive, is always uncertain and yet is necessarily coming and so therefore always demanding a response" (p. 249; cf. Anderson, 2010). In post-earthquake Nepal, the prospect of becoming a 'hydropower nation' authorized Nepali officials, the hydropower sector, financial institutions, and investors to act on its behalf – interceding to ensure its continued viability and value. Infused with the affective intensity of a prosperous hydropower future, the shareholder model coordinated collective forms of anticipation, investment, and solidarity in the wake of the earthquake. Facing threats to that imagined future, the state, the hydropower sector, and its investors sought to maintain the hopeful temporality of the 'not yet' (cf. Miyazaki, 2006; Bloch, 1986). Speculative dreams of becoming a 'hydropower nation' were given new life, and the map of Nepal's hydropower future was drawn anew over a seismically active landscape.

In recent years, several scholars have emphasized the precarity of infrastructures, reframing them as tentative achievements that may fail or fall to ruin (cf. Howe et al., 2016; Carse, 2014; Barry, 2013). Reflecting on the imaginative politics of infrastructure, Reeves (2016) has suggested that greater analytical attention to 'infrastructural hopes' is needed to examine the "articulation of material politics with diffuse elite and vernacular desires for a territorially secure future" (p. 711; cf. Harvey & Knox 2012). Amid a global resurgence in dam building (Zarfl et al., 2015) and struggles over resource sovereignty that may only intensify in the age of the Anthropocene (Folch, 2016),[10] there is good reason to reconsider the risks inherent to intensive hydropower development (Gergan, 2016; Drew, 2017) Huber et al., 2017. Recognizing these latent uncertainties is critical in the Himalaya, a region that is seismically active, vulnerable to climate change, and a source of drinking water for more than 1.4 billion people (cf. Immerzeel et al., 2010). As plans for intensive hydropower development in the Himalayan region continue to expand in scope and scale (cf. Drew, 2017; Gergan, 2016; Huber and Joshi, 2015; Alley et al., 2014) critical analyses that investigate "how and why dam construction continues to be projected as an orderly and safe activity, alongside the emphasized ideals of modernity, growth and clean/climate-friendly development despite a history of dam failures" (Huber et al., 2017: 51) are more important than ever.

And yet, in the wake of the 2015 earthquake, discussion about the imminent threat of seismic activity has been tellingly absent from an otherwise vigorous public discourse about the future of hydropower development in Nepal. Before and after the earthquake, hydropower developers and policy experts have repeatedly framed seismic risk as manageable and largely a financial problem, an issue that can be effectively mitigated by increased investment (cf. Rest et al., 2015; Butler and Rest, 2017). In this paper, I show that this state of affairs reflects a broader pattern, wherein "speculative capital operates as if the virtues of movement into valued futures are already known" (Adams et al., 2009. 251) – foregrounding particular conceptions of value, equity, and risk while obscuring or eliding others. The discursive ascendance of the shareholder model, understood as a 'capitalist technology of imagination' (Bear, 2015a) used to promote and coordinate speculative practices among a variety of different publics, is arguably the most telling indicator of this trend.

In the analysis that follows, I examine the ways that the 2015 earthquake prompted a reconfiguration of this imagined future and its economy of anticipation, showing how the persistent presence of the imagined hydropower future and the intensification of financial speculation in Nepal's hydropower sector has continued to preclude other ways of relating to uncertain and unknowable futures. Drawing on ethnographic research conducted before and after the 2015 earthquake, I highlight the ways that public discourse and official policy are increasingly configured in a way that "gives speculation the authority to act in the present" (Adams et al., 2009: 249). Focusing on the competing crisis narratives that emerged within the hydropower sector in the aftermath of the disaster, I describe the ways that different kinds of 'resource affect' (Weszkalnys, 2016) have shaped both public discourse and official policies designed to secure the hydropower future. Ultimately, I argue that concerns about seismic risk in the Nepalese Himalaya have been eclipsed by the logics of finance capital, a resurgence of speculative practices, nationalist rhetoric focused on energy sovereignty, and renewed commitments to the dream of becoming a 'hydropower nation'.

Making citizen-shareholders

In an interview conducted a few years after the Chilime IPO, Kul Man Ghising reflected on the success of the 'Chilime model' and the wealth his company had created for its local shareholders (ShareSansar, 2014):

> Chilime's model is singular in that it is a people-to-people partnership model, and not a Public-Private Partnership (PPP) model, which most other

companies have . . . Each of the households in project-affected areas, the poorest of the poor, now has shares worth NRs 25 to 30 lakh [~$25,000–30,000]. When they applied for the IPO the value of their entire property would have hardly stood at NRs 1 lakh [~$1,000]. This way Chilime has directly contributed to the GDP growth of the country by ensuring inclusive growth.

Highlighting strong community support in the Rasuwa district, he pointed to four other projects being developed by Chilime subsidiaries, and spoke of plans to build a high-elevation storage dam in the nearby Langtang Valley.[11] Following this process of capitalization to its logical conclusion he speculated that "within three years of these IPOs, all of the local shareholders of Chilime will become millionaires" (ShareSansar, 2014).[12] Pointing to Chilime and other infrastructure projects being developed in Rasuwa, locals often express pride in the ways that Rasuwa is 'becoming *bikasit* [developed]' (cf. Pigg, 1992) – as one enthusiastic shareholder once told me: "*Rasuwa pahila durgam, ahile hero ho*" ("Rasuwa was remote before, now it is a 'hero'").[13] Indeed, success stories about the Chilime project and the rapidity of social change in Rasuwa are well known across Nepal.

In the years since the Chilime offering, the concept of a 'people-public-private partnership' has proliferated rapidly across Nepal. Dozens of Nepali hydropower companies have filed for listing on the Nepal Stock Exchange, and each completed offering has been 'massively oversubscribed' (ShareSansar, 2016; Republica, 2015).[14] In response to the buzz, 'national priority projects' being developed by foreign contractors have also committed to conducting some kind of local share offering, reflecting an increasing effort to 'scale-up' the 'people-public-private partnership' model.[15] In short, citizen-financing and local investment in hydropower projects has become the 'new normal'. Popular excitement about 'getting shares' was aptly described in a *Kathmandu Post* op-ed from December 2014 (Tamot, 2014):

> These days, anyone and everyone on the streets and on social networks seems to be talking about local shares – from village folks directly hit by such dams and their neighbors to city dwellers looking for lucrative profit-making ventures, from businessmen and politicians to our migrant brothers and sisters toiling in distant lands.

The popularization of the shareholder model in Nepal is partially connected to an increasing emphasis on 'sharing the benefits' of hydropower development (cf. Shrestha et al., 2016) – a discursive trend that refers both to globally circulating ideas about 'sustainable' hydropower development and to the specific history of Nepal's own 'constructive dialogue on dams and development' (cf. Dixit and Gyawali, 2010). Conceived of as a corrective to extractive development practices, benefit-sharing programs seek to enroll 'project-affected populations' in a shared agenda of improvement, creating new kinds of stakeholder relations and corporate subjectivities. And yet, while concepts like 'benefit-sharing' and 'corporate social responsibility' are clearly and formulaically defined in theory, in practice they are continually *negotiated* and often *redefined* through multi-scalar contestations among differently positioned stakeholders (Lord, 2016; Shrestha et al., 2016). Similarly, these differently positioned stakeholders have formed a variety of opinions about the financial, political, and affective functions of 'the shareholder model'.

Viewed from a normative economic perspective, the advent of benefit-sharing and profit-sharing programs might contrast favorably with past patterns of displacement and practices focused simply on mitigating harms – signaling a democratic and 'sustainable' turn in the history of hydropower development. In this light, increasing focus on 'local benefits' might seem like a promising effort to rebalance the historically uneven distribution of costs and benefits between urban centers and rural peripheries.[16] From the pragmatic perspective of a Nepali financier, the shareholder model appears to be a key technology for mobilizing domestic capital – decreasing dependence on donor institutions or foreign development banks, and rapidly addressing increased energy scarcity. Importantly, however, it should also be recognized that Nepalis purchasing equity shares in the imagined hydropower future are also sharing *costs* and *risks*.

When hydropower developers talk about share offerings and benefit-sharing programs, they frequently highlight the ways these programs help generate a 'social license to operate' for their companies (Shrestha et al., 2016: 15). Similarly, they often emphasize the ways that 'getting shares' can foster a 'feeling of owner-ship' within project-affected communities. Speaking about the ways that local investment helps align incentives, one industry representative told me: "*mero aaphno paisa die bhane tyo project disturb gardaina*". ("If I give my own money [by buying shares in the project] then I won't disturb that project".) He argued that the people of Rasuwa had 'learned these things from the Chilime project' and others would too: "if they see the project stopped for twenty-four hours, then they see their share prices go down". In short, he was optimistic that local shareholders would begin to recognize the logics of finance capital, and that they would act in their own best interests.

Seen from a critical perspective, however, these same programs might also be construed as technologies for manufacturing consent with 'project-affected communities', as adaptations that warp or compromise legally-mandated processes of stakeholder consultation and commitments to 'free prior and informed consent'. Similarly, one might interpret company efforts to create local investors as a means of rendering local populations 'feasible' (Golub, 2014: 20), enacting ideas of corporate social responsibility that also help to safeguard capital accumulation among unruly publics (cf. Rajak 2011; Welker, 2014; Barry, 2013). As I hinted earlier, one could also view these programs as evidence of an emerging interest in practices of 'expectation management' wherein state and corporate actors attempt to translate popular forms of 'resource affect' into capital: "substituting the wrong, futile, or excessive kind of hopes for the right ones" (Weszkalnys, 2016: 142). In an age of public austerity, these attempts to remake

Speculation and seismicity   173

infrastructural publics into popular speculators who bear new kinds of risk may reflect a broader trend (cf. Bear, 2015b).[17]

On a different note, one could also argue that the discursive ascendancy of the shareholder model in Nepal suggests an ethical re-centering of the economy of anticipation around the rights of capital and the protection of future shareholder value. While the profit-oriented agenda of 'hydraulic bureaucracies' was previously obscured by a developmentalist rhetoric of impartiality and the greater civic good (cf. Molle et al., 2009), the shareholder model signals a turn toward a 'market-embedded morality' (Shamir, 2008) wherein the generation and distribution of profit is considered a 'common good'. The extension of the financial logics of shareholder capitalism – wherein "shareholders are typically assumed to be alike in their desire for maximized risk-adjusted financial returns; their personal moral beliefs are held separate from their actions as shareholders" (Welker and Wood, 2011: 60) – risks collapsing the complexities of citizenship and ethical obligations to 'project-affected people' into a single subject category.[18] Exploring the contours of contemporary 'shareholder alienation' Welker and Wood (2011) have argued against the singularity and normativity of these market-based logics, highlighting the inherent relationality of shareholders and potential for shareholder activism. They argue that "rather than unified and coherent possessive individuals with transparent desires, shareholders are fractured, multiple, and composite" (p. 65). Similarly, the shareholders of Nepal are neither homogenous, nor apolitical, nor passive.

In recent years, an array of differently situated Nepalis have demanded the right to purchase shares in the hydropower projects that impact their lives, using their official status as *ayojanale prabhabit manchhe* ('project-affected people') to demand recognition and rights. In one case, when a landslide damaged the facilities of the Bhotekoshi hydropower project in 2014, a group of local leaders prevented the hydropower company from repairing the damage and demanded that locals in the 'project-affected area' be given shares. In another case, laborers at the Upper Tamakoshi Hydropower Project organized a strike to demand shares, arguing: "We have invested our sweat, our blood, our lives in the project tunnels. We are the most-affected people. Why shouldn't we be able to invest our money?" (Lord, 2016). In response to these encounters, some policymakers in Kathmandu cried foul, highlighting that the financial costs of these strikes would be passed on to their fellow shareholders, and claiming that no citizen has the right to obstruct the development of the country. Others began referring to a 'shares virus' that might scare away foreign investors and ultimately jeopardize the making of the hydropower nation. Alternatively, one policy expert prophetically suggested that "some bugs, if they don't kill you, make you stronger" (Tamot, 2014: 1).

The earthquake and the hydropower nation

In April of 2015, Nepalis across the country were waiting expectantly for the initial public offering of the Upper Tamakoshi project – a 456 MW 'national priority' project conducting the largest IPO in the history of the Nepal Stock

Exchange, valued at NRs 5.18 billion, or roughly USD $50 million. Over 400,000 Nepalis queued across Nepal to file the official paperwork to be registered to purchase shares (Republica, 2015). As the IPO date grew closer, project laborers and local 'project-affected populations' organized strikes, demanding the right to purchase *more* shares. National newspapers and the evening news were saturated with 'resource affect'. On April 17, a group of interested parties (myself included) gathered in Kathmandu for a workshop called "The Shares Question: Financial Equity and Hydropower Development in Nepal".[19] The economy of anticipation was in especially robust form.

And then, at 11:56 AM on April 25, 2015, while the modalities and allocations of the Upper Tamakoshi offering were still being debated, Nepal experienced a 7.8 magnitude earthquake that killed almost 9,000 people, destroyed over 600,000 homes, and caused significant damage to infrastructures (Government of Nepal, 2015). The earthquake and its aftershocks also had significant effects on the hydropower sector and the national grid: seventeen different hydropower facilities, representing 15% of the country's installed generation capacity, were 'severely damaged', leaving 600,000 households without electricity (Government of Nepal, 2015). There were no dam failures – arguably, because the larger reservoir-type high dams had not yet been built.[20] Dozens of hydropower projects still under construction across Nepal were damaged or delayed, incurring new costs and exacerbating energy scarcity (The Kathmandu Post, 2015). The work of making the hydropower nation was halted; dreams of the hydropower future were momentarily deferred.

Speaking to the uncertainties of this moment, Gagan Thapa, a Member of Parliament and the acting Chair of the Parliamentary Committee on Agriculture and Water Resources, provocatively co-authored a *New York Times* op-ed that questioned the wisdom of the dominant hydropower development paradigm. Advocating for a rethinking of energy security and a diversification of Nepal's energy mix, the authors called for greater attention to the seismic and ecological vulnerabilities of large-scale hydropower infrastructures (Thapa and Shrestha, 2015: 1):

> Recent events raise many serious questions for a country whose economy and energy policy is hydropower-based: Why are modern plants, many constructed in just the past few years, unable to withstand disasters that are not the worst Nepal has been expecting? Are we paying adequate attention to the location of the sites where these plants are being built, and factoring in vulnerabilities? What is the quality of construction of these projects? Are changes in landscapes and ecosystems (deforestation, for example) putting our hydro projects at greater risks? Climate change aside, can Nepal rely on its built and planned hydro infrastructure given the inevitable seismic activity in the Himalayas?

This line of questioning, alongside other commentaries focused on 'the damage done and the dams to come' (Hilton, 2015; Rest et al., 2015) indicated the possibility of critically rethinking the shape of the imagined hydropower future. This reflective moment, however, would eventually fade.

In July of 2015, Nepal's hydropower community gathered in Kathmandu for an event titled "The Impact of Disaster on Hydropower Development in Nepal". The program began with admittances, such as the statement by Radesh Pant, the CEO of the Investment Board of Nepal, who said: "We all knew that we were in an earthquake-prone zone, but you don't really know until you have experienced it." Dr. Sandip Shah, head of the Norwegian developer SN Power, argued comprehensive risk assessment must begin very early in project planning and told the crowd, "We can no longer be complacent." Several speakers mentioned the need for better coordination and strategic planning, some highlighting the need for strategic 'cumulative impact assessments' in the post-earthquake era. Dr. Bishal Nath Uprety of the Nepal Academy of Science and Technology said quite directly: "This is the time to rethink."[21] Nepali seismologists and Japanese experts in earthquake engineering gave detailed presentations. Everyone present seemed to acknowledge the enormity of the unknowns. And yet, this tentative consensus would soon fade, as familiar market-oriented arguments resurfaced.

Taking the podium to provide an update on earthquake damage to the Upper Tamakoshi project,[22] the project's managing director Bigyan Shrestha began his remarks with a half-joke: "I think the status of the Upper Tamakoshi is in everybody's interest, because right now we have more than two lakhs [200,000+] shareholders." The roomful of representatives from the hydropower industry laughed nervously.

The Upper Tamakoshi project had fared relatively well during the earthquake, he reported. Though the foundation of the diversion dam settled 20 cm into the riverbed, there was no major damage to the subsurface structures – a network of tunnels roughly 25 km in length, including the massive powerhouse cavern 1.4 km underground. Construction facilities and employee housing units were damaged by landslides and some equipment had been damaged by flooding and erosion, but there were few outright collapses. The most debilitating damages were inflicted on the road system, as recurring landslides obstructed movement to and from the project site and the main access road to the headworks site at Lamabagar remained blocked for the duration of the monsoon. Though casualties occurred in many of the surrounding villages, no one was killed at the project site.[23] No project employees were significantly injured; all project employees (including foreign contractors) were evacuated as soon as possible; and the Nepali workforce was given leave to return home to help with relief and recovery work. While construction had not yet resumed due to access issues, things had stabilized, and the project sought to improve its planning.

With this report, Shrestha provided a detailed and admirably transparent account of earthquake-related damages and post-disaster response activities. While Shrestha has elswhere acknowledged the threat of seismic risk,[24] the normative tone of this report reflected a common industry assertion that geological risks are manageable, reinforcing the belief that seismic activity was "a calculable risk that could be managed through technological intervention" (Rest et al.,

2015: 2). As the event continued, the conversation gradually turned from seismic risk to financial risk. Classic questions of accountability and financial viability returned: Who exactly will assume the responsibility for disaster risk reduction measures or cumulative impact assessments? Who will bear the costs? Reflecting the encompassing nature of financial logics, one of the civil society representatives present asked, "Are investors interested in paying for these programs?"

Riffing on this question, representatives from the Nepal Hydropower Association (NHA) expressed concern over the 'inflated risk perception of investors' after the earthquake, while another private developer warned that the increased costs of seismic risk preparedness and project monitoring could disproportionately affect the financial viability of smaller projects. Another speaker encapsulated the mood by placing an emphasis on corporate messaging: "We have to communicate to investors that disaster management and preparedness is important to Nepal." Slowly, the conversation shifted from concrete recommendations for policy reform and seismic safeguards, back to the management of financial risks. Very little time was spent discussing the risks of dam failure attached to the large reservoir projects planned for the future, while a great deal of time was dedicated to topics of financial liability, the need for subsidized loans to developers struggling with post-earthquake repairs, and efforts needed to promote investor confidence. Gradually, future risks were hierarchically sorted, priced, and incorporated into a thin model of the future perfect (cf. Guyer, 2007).[25] Participants sought to regain lost momentum and restore the promise inherent within this economy of anticipation, and conversation inevitably shifted back to the importance of retelling the story that 'Nepal is [still] open for business'.

Discussing the ways that Japanese financiers, operating at the speculative edges of capitalism, interpreted crisis and collapse in the 1990s, Miyazaki (2006) that a hopeful "redeployment of logicality" helped Japanese traders to reorient themselves (p. 160). Crisis initially prompted these traders to question the epistemic value of their financial models, but "hope surfaced repeatedly in the uses of these ideas and tools despite their repeated failures, and perhaps even because of these failures" (p. 151).[26] In short, the financial crisis that threatened their faith in market principles was not the end, but a moment for reimagining the *telos* of capitalism and the space of the 'not yet' (Bloch, 1986) in which it thrives.

Confronted with the crisis of the 2015 earthquake, which threatened the feasibility of dam projects and highlighted the possibility of future disasters, representatives of Nepal's hydropower sector seem to have performed a similarly hopeful reorientation – redeploying existing financial logics and discursive methods to keep their hopes of building a 'hydropower nation' alive. These replications and reconfigurations were apparent in the discursive trajectory of this post-earthquake hydropower sector event, where a "reprise of method" allowed hope to spring forth anew (Guyer, 2017: 158): as a reassertion of dominant financial narratives, only slightly retooled.[27] In response to the palpable uncertainties and the concerns about *closure* that prompted this meeting, the discourse on hydropower development recrystallized in predictable formations, keeping the

Speculation and seismicity 177

hydropower frontier and its correspondent future *open*. In the course of a single two-hour event, one could observe the way that "hope surfaces through the interplay of openness and closure" (Miyazaki, 2017: 26). And as the aftermath continued, this pattern repeatedly emerged.

Energy emergency and 'the people's investment'

In September 2015, the conversation about Nepal's energy future was complicated by an 'unofficial blockade' along the Nepal-India border in response to the rapid promulgation of a contentious new constitution in August (Sangraula, 2015).[28] Lasting four months, this blockade effectively shut down all legal commerce over the Nepal-India border, highlighting Nepal's chronic energy insecurity. Because Nepal depends on India for nearly all of its fuel imports, this blockade precipitated a 'fuel crisis' with serious material, financial, and geopolitical consequences. Families struggling to cope with the earthquake damage and facing up to 16 hours a day of scheduled blackouts known as "load shedding", millions of Nepalis had to make do without petroleum or cooking fuel. Post-earthquake recovery and reconstruction work was delayed significantly, and 'national priority' infrastructure projects came to a halt (Pandey, 2016; Pangeni, 2015). Citing economic losses valued at more than USD $1 billion, several public officials and industry leaders described the shutdown as an economic disaster "worse than the earthquake" (Pattison, 2015).

Responding to the crisis, policymakers competed to present strategies for building a secure energy future and decreasing energy dependence on India. Public dialogue was saturated with nationalist polemics about reclaiming Nepal's 'energy independence' and affective rhetoric invoking the possibility of future threats to energy sovereignty. The private sector clamored for the declaration of an 'energy emergency' that would lift all existing barriers to rapid development in the Nepalese hydropower sector (Pandey, 2016). As the lights flickered in Kathmandu and people began burning wood salvaged from earthquake debris, the hopeful call "to keep an entire generation from growing up in the dark" (Lord, 2014: 112) resurfaced in a new context. As the blockade continued, interest in rethinking Nepal's path to energy security evaporated, and the longstanding goal of rapid hydropower development was re-inscribed in the national psyche.

In February of 2016, the government of Nepal declared a National Energy Crisis Reduction and Electricity Development Decade and formulated an official ten-year plan organized around two highly aspirational goals: providing electricity to every household and helping every Nepali become a shareholder (Government of Nepal, 2016) Revealingly, the official motto of this plan was '*Nepalko Paani, Janatako Lagaani*' ('Nepal's Water, the People's Investment'). Arising from a fusion of speculative exuberance and nationalist concerns, this slogan reflected both a reframing of familiar articulations that "not one drop should flow beyond Nepal's borders without creating wealth" and a significant scaling-up of the shareholder model.[29]

178 *Austin Lord*

Speaking about the strategic use of crisis, Milton Friedman once suggested that:

> Only a crisis – actual or perceived – produces real change. When that crisis occurs, the actions that are taken depend on the ideas that are lying around. That, I believe, is our basic function: to develop alternatives to existing policies, to keep them alive and available until the politically impossible becomes politically inevitable.
>
> (Klein, 2007: 7)

In many ways, it seems that the hydropower sector and their state allies seem to have pursued this opportunistic strategy: leveraging the urgencies of the fuel crisis and earthquake-related damage to enact a set of long-desired free market reforms. In this sense, these crises helped to create spaces of exception (cf. Ong, 2006) required to protect the imagined hydropower future – using the logics of shareholder capitalism to authorize state interventions justified by a desire to protect future shareholder value and secure returns on investment.[30] While the hydropower sector has advocated for the declaration of an 'energy emergency' several times in the past (Lord, 2014: 115), the earthquakes and the crises that followed created a discursive opportunity for these demands to be translated into legislative action.

Meanwhile, in spite of the earthquake, the fuel crisis, and ongoing political volatility, nearly a dozen other hydropower companies have filed new offerings with the Nepal Stock Exchange (ShareSansar, 2016). Tellingly, just six months after the earthquake and *during* the blockade, the government-endorsed Hydropower Investment Development Company Limited (HIDCL) conducted a wildly successful initial public offering. While the company sought to raise NRs 20,00,00,000 (~USD $20 million), the company received applications worth an astounding USD $350 million from over 300,000 Nepali applicants. The financial media reported triumphantly that "this IPO has seen overwhelming response from all kinds of investors even when the country is going through one of the biggest crisis our generation has seen". (ShareSansar, 2015).[31] At an event hosted by the International Finance Corporation in December 2016, Niraj Giri, the head of the Securities Exchange Board of Nepal, reported that the Nepali general public had so far invested a total of NRs 11.55 billion (~$110 million) in hydropower.[32] Echoing a hopeful forecast commonly repeated among industry and state actors, he suggested that a great deal more domestic capital was available and waiting.

Exploring the ways that Spanish hydropolitics were shaped by narratives of national 'regeneration', Swyngedouw (2015) highlighted the value of an analytic focused on the politics of convocation and "on who or what does the summoning" within a landscape where multiple choreographies and assemblages are possible (p. 227). In the wake of the earthquake and the fuel crisis, the government of Nepal has summoned the citizens of the imagined 'hydropower nation' to participate in a 'people-public-private partnership' that will secure the hydropower future. Using the speculative phrasing '*Nepalko Paani, Janatako Lagaani*' the state seeks to leverage the contrast between two divergent futures. On one hand, the idea of an 'energy emergency' invokes fear of future resource scarcity and pre-empts potential threats to energy sovereignty. On the other, the inclusive and moralizing rhetoric of 'the shareholder model' is used to choreograph popular investment, invoking visions of a prosperous hydropower future perfect. The use of these 'capitalist technologies of imagination' facilitate new patterns of convocation[33] – as the concept of energy sovereignty is reconfigured to promote the rights of citizen-shareholders with claims to the future value of Nepal's water. Emboldened by financial logics, nationalist dreams of sovereign hydropower wealth, and perennial hopes for infrastructural security, the state has redoubled its efforts to support rapid hydropower development.

In short, throughout and despite a series of recent crises, 'the shareholder model' of hydropower development has maintained its discursive and financial momentum. In fact, its powers of convocation seem only to have increased. Recent changes in hydropower policy have realigned efforts to securitize the hydropower frontier with the financial securitization of the hydropower future – providing further evidence that "the present is governed, at almost every scale, as if the future is what matters most" (Adams et al., 2009: 248).

Amnesia, ignorance, and the open future

It is often presumed that the ruptures and instabilities of disasters will threaten the status quo: that new discursive spaces will open, creating opportunities for critical reinterpretations of past certainties and the reevaluation of normative futures. When the earthquake struck Nepal on April 25, 2015, it exposed latent patterns of vulnerability, seismic uncertainty, and systemic risk, introducing a moment to 'rethink' Nepal's imagined energy future (cf. Thapa and Shrestha, 2015; Rest et al., 2015). And yet, my analysis indicates that this resurfacing of seismic and geopolitical risks has not prompted a meaningful re-evaluation Nepal's energy future – that the imagined hydropower future was, in a deeply affective sense, 'too big to fail' (cf. Lakoff, 2010).[34] Concerns about seismic risk and geohazards have been crowded out by recurring anxieties about energy scarcity, national sovereignty, and the enthusiastic flow of capital toward Nepal's imagined hydropower frontier. Renewed forecasts of an abundant hydropower future and affective investments in energy security have sucked the political and financial oxygen out of the room, and the Nepalese state seems to have adopted an all too common post-disaster strategy of 'planning to forget' (Simpson, 2013: 245). Given the scale of seismic uncertainties in the Himalayan region and the frequent incidence of cataclysmic geohazards (cf. Kargel et al., 2016) one has to wonder: do these claims represent a form of infrastructural hubris (Lord 2017)?

In a recent paper, Huber et al. (2017) consider the ways in which the threats of dam failure is rendered 'unthinkable' and repeatedly omitted from debates about hydropower development. Highlighting a politics of amnesia, they call for greater investigation into "how risk acceptability is negotiated, through which

180 *Austin Lord*

processes risk materializes in the form of disaster, and how the state creates disaster vulnerabilities" (Huber et al., 2017: 54). In post-earthquake Nepal, the continued success of the shareholder model, in spite of significant uncertainties and myriad risks, highlights one kind of answer – testifying to the affective power of infrastructural futures and the inertia of speculative financial logics. In Nepal, the primacy of speculative logics and the rights of capital effectively warp dialogue about disaster risk reduction, glossing over uncertainty and reproducing a kind of '*strategic ignorance*' (McGoey, 2012). For the state and the hydropower sector, I argue, "ignorance serves as a productive asset, helping individuals and institutions to command resources, deny liability in the aftermath of crises, and to assert expertise in the face of unpredictable outcomes" (p. 553). As Bear et al. (2015), have suggested: "speculation fuels, and is fueled by, a heightened sense of anticipation in which routines of calculation are often suspended" (p. 388). In the Nepalese Himalaya, where abundant hydropower resources and seismic risk compete for attention (Lord 2017), some risks are capitalized and others remain strategically unknown.

In the same way, dreams of making a 'hydropower nation' turn thousands of Nepalis living along the unstable hydropower frontier into "unimagined communities" – people who are "physically unsettled and imaginatively displaced, evacuated from place and time and thus uncoupled from the idea of a national future and a national memory" (Nixon, 2010: 62). Teleological assumptions about the hydropower future allow the state to ignore the existence of vulnerable communities, such as the people who have been or will be displaced by dams, those living in areas where the process of hydropower construction intensifies landslide risk, or the potential victims of a possible dam failure. These Nepalis are *not* the idealized citizen-shareholders of the imagined 'hydropower nation' whose electricity needs and claims to future value lend moral force to the speculative ambitions of the state and the hydropower sector, but peoples whose presences "inconvenience or disturb the implied trajectory of a unitary national ascent" (ibid). Here, anticipation facilitates a project of erasure, in which the disastrous possibilities of the past, present, and future are obscured by interventions justified as investments in the security, productivity, and collective well-being (cf. Klein, 2007; Adams et al., 2009).

At the same time, the rise of the shareholder model implies another kind of 'imaginative displacement' that is based on the *inclusion of* Nepal's unimagined communities. As this 'capitalist technology of imagination' proliferates, it seeks to redefine 'project-affected populations' and displaced peoples as 'local shareholders' who will share in the potential benefits of the speculative hydropower future. In this way, it creates both 'popularist speculators' (Bear, 2015a) who help finance the imagined hydropower future and the citizen-shareholders whose needs enhance the authority of the Nepalese state. This kind of imaginative displacement is operative in the state slogan '*Nepalko Paani, Janatako Lagaani*': a nationalist prophecy that invokes both moral claims to a sovereign future and popular 'resource affect' (Weszkalnys, 2016) among Nepal's imagined infrastructural publics. Emerging as a source of capitalist hope in the wake of recent crises

Speculation and seismicity   181

(cf. Miyazaki, 2006), the dream of a national 'people-public-private-partnership' is being used to coordinate renewed efforts toward the realization of a dream deferred.

All told, the April 25, 2015 earthquake has *not* interrupted the hopeful project of becoming a 'hydropower nation'. While the disaster made the material precarity of the imagined hydropower future immediately apparent, these frictions were eventually translated into a form of political *traction*, as desires for the hydropower future gained discursive *momentum*. In some ways, this is not surprising, for as Tsing (2005) has argued in her classic study of the encounters that shape spectacular resource frontiers: "friction is not just about slowing things down" (p. 6).[35] The geophysical risks that surfaced during the disaster have been priced into expert financial models and politically metabolized within the broader economy of anticipation.[36] Seismic risk has been eclipsed by other future-oriented concerns, and the hydropower sector has returned to its business with a renewed sense of urgency and anticipation. Hope springs eternal along Nepal's hydropower frontier.

Nepal's hydropower future, however, remains deeply uncertain and far from over-determined. As several scholars have recently highlighted, infrastructures and their infrastructural publics are constantly being made and unmade, and the fact that "material entities are not easily subdued by prognostic practices" (Mathews and Barnes, 2016: 21).[37] Speaking to the uncertain temporality of infrastructural promise Gupta (2015) recently remarked that "the bridge to the future is always under erasure, and we do not know where it will lead" (p. 1). This ambiguity is particularly intense in Nepal, where infrastructural ambitions must contend with 'uncertainty on a Himalayan scale' (cf. Thompson et al., 2007; cf. Gergan, 2016; Drew, 2017). Attempts to harness the power of rivers have always been a struggle for as White (1995) so simply stated, while writing about the entropic properties of the Columbia River: "the river has purposes of its own which do not readily yield to desires to maximize profit" (p. 112). Climactic variability that threatens to increase erosion, raise the risk of landslides or glacial lake outburst floods, and intensify monsoonal flooding raises another set of unknowns in the Himalaya. Lastly, and perhaps most importantly, while the timing of seismic events cannot be predicted, it is certain that earthquakes will continue to occur in the Nepalese Himalaya (Kargel et al., 2016). Because the potential energies capable of disrupting imagined hydropower futures inhere within the present, the future remains radically open.

Notes

1 'Local shares' are equity shares sold to local 'project-affected populations' that are publicly traded on the Nepal Stock Exchange (NEPSE). As a result of legal disputes, the Chilime Hydropower Company conducted its local shares offering in 2010, two years after it completed its initial public offering (IPO) of shares to the general public. For more details on the history of contestation surrounding the Chilime offering, see Lord (2016: 154–156).
2 Tellingly, 31,123 people purchased shares in the Chilime local offering in 2010, whereas only 23,675 people voted in the district of Rasuwa during the national elections in 2013 (Lord, 2016: 146).

3 As Hetherington (2016) has suggested: "the tense of infrastructure, like any development project, is the future perfect, an anticipatory state around which different subjects gather their promises and aspirations" (p. 40). Gyawali (2006) describes the deferral of Nepal's hydropower future perfect as follows: "Nepal's barons of hydro-policy have long promised a bonanza from the waters cascading down the Himalaya, if and when developed into electricity from high dams and exported into India. Water, and hydroelectric potential in particular, is described in many a ministerial speech as the country's 'only resource': a resource, moreover, of such magnitude that Nepal is 'second in the world in hydropower potential . . . [and yet] the hype of the last half-century – and a water development policy stemming from that hype – has left Nepal in a sorry state at the beginning of the new millennium" (p. 61).

4 The 'making of the hydropower nation' (Lord, 2014) has been repeatedly troubled by political volatility, shaky investor confidence, and social conflict along Nepal's imagined hydropower frontier. See also Pandey (1996), Forbes (1999), Gyawali (2003), Rest (2012), and Butler (2016) on the history of hydropower-related ambition and developmental failure in Nepal's hydropower sector.

5 As of October 2017, eighteen hydropower companies are publicly listed on the Nepal Stock Exchange and another thirty-five companies have filed to conduct initial public offerings (ShareSansar, 2017).

6 See also Sneath et al. (2009) and Appadurai (2011) for a discussion of the ways that 'technologies of imagination' are used to shape imaginative capacities and notions of futurity.

7 Weszkalnys (2016) suggests that "resource affect may take a variety of forms: euphoria, excitement, aggression, doubt, trepidation, frustration, disillusionment, and so on. These regularly emerge, successively or alongside each other, in contexts of resource prospecting and extraction" (p. 128).

8 Writing about the material and affective experience of infrastructure, Knox also suggests that politics "emerges and is reworked through affective engagements with the material arrangements of the worlds in which people live" (p. 375).

9 The earthquake of April 25, 2015 and the aftershocks that followed caused 8,964 casualties in Nepal, leaving almost 22,000 people injured and an estimated 3.5 million homeless (Post Disaster Needs Assessment, Government of Nepal, 2015). See below for a more comprehensive account of the infrastructural damage caused by the seismic event.

10 In a recent paper, Folch (2016) explores Paraguayan claims to 'hydroelectric sovereignty' in Paraguay, where nationalist actors seek self-determinataion vis-á-vis Brazil. As Folch describes, this claim to resource sovereignty is articulated in terms of a territorial claim on the future, a claim to resource flows amid the uncertainty of the Anthropocene.

11 The Rasuwaghadi (111 MW), Sanjen (42.5 MW), and Upper Sanjen (14.5 MW) hydropower projects are all expected to complete construction and begin commercial operation in Rasuwa before the end of 2018. Surprisingly, the planning process for the Langtang Storage Project (~350 MW) also continues, despite significant damage in this area during the 2015 earthquake, and significant risk of future geohazards.

12 Critically, this is not an exaggeration. The stock price of Chilime shares did rise considerably over time (once trading at about twenty times the initial issue value), which resulted in the distribution of dividends and even bonus shares to local shareholders in Rasuwa (the stock price has since trebled, but remains around ten times the offer price). However, this type of share price performance will be exceedingly difficult to replicate, and many say that it has created 'unrealistic expectations' for future hydropower share offerings.

13 This trend illustrates the ways in which hydropower development taps into socially-constructed aspirations of 'becoming *bikasit* [developed people]' that circulate widely in Nepal, and the ways in which proximity to the material comforts of *bikas* (development) is a common marker of social distinction (Pigg, 1992; Rest, 2012).

14 In the wake of legal disputes over the structure of the Chilime shares offering, the Securities Exchange Board of Nepal (SEBON) created a rule that requires all hydropower companies seeking to go public on the Nepal Stock Exchange (NEPSE) to reserve 10 percent of shares offered for people residing in the 'project-affected area' determined in the

Speculation and seismicity 183

environment impact assessment. For example, the IPOs for the smaller Ridi Hydropower Company (RHCL) and the Barun Hydropower Company (BHCL) were oversubscribed by sixty and sixty-one times, respectively (Republica, 2015).

15 For example, the project developers of the Upper Tamakoshi (456 MW), Upper Trishuli 1 (216 MW), Arun 3 (900 MW), Upper Karnali (900 MW), and Budhi Gandaki (1200 MW) hydropower projects have all committed to conducting local share offerings. Because some of these projects are being developed by privately-held foreign companies, the complexities of the offering modality these companies will use is still being determined – the fact that they are going to such lengths to make a 'local share offering' possible is itself incredibly telling of the discursive momentum of this model.

16 See Pandey (1996) for an early analysis of center-periphery relations in the hydropower sector and the possibility of 'local benefits from hydropower development'. See also Forbes (1999) on the scalar questions of defining locality in the context of debates over the Arun III project.

17 In recent years, citizen-investment schemes have been used to finance hydropower infrastructure and nation-building in a few other places around the world, such as Tajikistan (cf. Menga, 2015: 486). These programs, however, often resemble *compulsory* state-mandated investment schemes, whereas the public interest in purchasing shares is entirely *voluntary* within Nepal.

18 This simplification risks legitimating an ethical regime governed by Milton Freidman's famous neoliberal dictum that "the social responsibility of business is to increase its profits" (Welker and Wood, 2011: 62).

19 This event, which I co-organized with colleagues from the Niti Foundation and Social Science Baha, convened a variety of representatives from the hydropower sector, government, civil society, and local concerned groups to discuss the financial, social, and political issues related to 'the shareholder model'.

20 Importantly, a crack *did* emerge in the dam of the single reservoir-type hydropower project in operation at the time of the earthquake; the Kulekhani I (60 MW) *did* suffer an 8-meter crack in its dam – however, there was no leakage because water levels were low during the late dry season.

21 Later in the meeting, Dr. Uprety also suggested that: "the Government has to provide a mandate that some percentage of the project budget must be spent on disaster management, even one half to one percent".

22 The 456 MW Upper Tamakoshi project had been affected by both the April 25th earthquake and the 7.3 magnitude aftershock that occurred on May 12, with an epicenter less than 15 kilometers away from the project site.

23 Though, sadly and ironically, a team of six people were killed just upstream from the project site while conducting an assessment for the planned 160 MW Lapche Khola Hydropower Project, a different hydropower project located further into the imagined hydropower future.

24 Bigyan Shrestha would later highlight the need for greater disaster planning in a published interview: "Before [the] April 25 earthquake, no one had made an exact calculation on how earthquakes would damage these projects. Nepal lies in a seismically active zone, particularly the Himalayas and hills, we need to seriously study seismic vulnerabilities of the region before taking the projects" (Shrestha, 2015).

25 Talking about temporal reconfiguration and the increasing displacement of the near future, Guyer suggests that the "spaces opened up [by this temporal reorientation] offer innovative extrapolation from some vantage points and block any cumulative momentum from others" (Guyer, 2007: 416).

26 Reflecting on the temporal qualities of hope, Miyazaki has suggested that hope emerges from a "temporal reorientation of knowledge and its resulting replication of past hope in the present" (2004: 130) and that "hope always contains within itself known forms replicated into not-yet-known terrains" (2017: 26).

27 Again, this is not to say that these are not valid concerns, or that companies like the Upper Tamakoshi are not responding in appropriate ways to post-earthquake needs (rather, as

184   *Austin Lord*

I hinted above, they are trying), but only that the primacy of financial narratives persists and that this event reflected these concerns.

28 In the wake of the earthquake, the protracted process of drafting a new constitution was fast-tracked, resulting in the promulgation of a contentious document that left several critical and contentious issues unresolved. The release of the draft constitution and ongoing debate over its content triggered a wave of political protests and violence in the Tarai region of southern Nepal in August of 2015. The Nepalese state mobilized security forces to quell dissent, but further violence from both sides provoked mass public protests. In reaction to these disturbances, the government of India formally and informally entreated the government of Nepal to address the demands of the agitating parties. When the government of Nepal failed to respond as desired, an 'unofficial blockade' materialized at several major transit points along the Nepal-India border. However, neither India nor the agitating political parties of the plains accepted full responsibility for the blockade – and so attributions of responsibility for the 'unofficial blockade' remain a subject of intense debate and research.

29 This phrase, publicly declared by a political leader participating in a 2013 event called the Power Summit, reframes a globally circulating mantra of nationalist water resource management in financial terms (Lord, 2016: 150).

30 Within the policy framework of the 'energy emergency', environmental issues and social concerns are interpreted as 'barriers to investment' and subordinated to the desire to control operational risks and secure the hydropower frontier as a space open for investment. These exceptional new policies include: fast-tracking the process for environmental review, privatizing contracts for transmission line construction, establishing a 'special tribunal' that will oversee contentious processes of land acquisition, classifying hydropower sites as 'restricted areas', and allowing private contractors to deploy state security forces in response to "obstruction from political parties and locals", and (vaguely defined) "allowing quake hit projects to exercise force majeure" to ensure that these projects are developed without further delay (Kathmandu Post, 19 February 2016).

31 Importantly, HIDCL is not technically a hydropower development company but a government-backed financial institution that seeks to develop and manage a portfolio of different hydropower projects – investors were buying into a fund that did not yet have real assets, just development licenses and 'paid up capital'. Massive investment in the HIDCL therefore adds an additional layer of speculation.

32 This event focused explicitly on 'Local Shares in Hydropower' and was part of a multi-year 'technical workshop series' being organized by the International Finance Corporation and USAID in Nepal.

33 Because practices of speculation and forecast often resemble a form of 'capitalist divination' (Bear, 2015: 411; cf. Appadurai, 2011), the language of convocation and invocation is particularly apt.

34 In the wake of the BP Oil Spill, Lakoff (2010) used this phrase, which initially referred to banks rescued with federal funding during the 2008 financial crisis, to discuss the denial of systemic risks implied by energy production in the Gulf of Mexico. Focusing on nationalist hydropower development schemes in Tajikistan, Menga (2015) has described a similar pattern of post-disaster persistence, arguing that "the idea of building the Rogun [Dam] was already too well established in the minds of bureaucrats to be washed away with the flood" (p. 483).

35 Focusing on Indonesia, Tsing (2005) uses the metaphor of *friction* to analyze the encounter between universalizing market logics and locally situated realities, arguing that frictions inflect motion and alter trajectories, resisting in some cases and enabling in others. "Friction is not a synonym for resistance. Hegemony is made as well as unmade with friction" (p. 6). An earthquake is a transfer of tectonic frictions into an incredibly destructive force, but this energy is translated into other forms of power within the political economy of disaster.

36 Similarly, Cross (2015) has argued that: "Just as crisis is the ground on which capitalism regenerates itself, so too [sic] unrealized futures constitute the fertile ground on which new hopes, dreams, and desires are produced" (p. 435; cf. Cross, 2014).

37 Several scholars have shown how infrastructures are tentative and precarious achievements that exist in tension with "demanding environments" (Carse, 2014: 180) requiring significant energy and coordination to maintain (cf. Reeves, 2016; Howe et al., 2016; Knox, 2017) in the face of unruly or 'inhuman nature' (cf. Clark 2011). Focusing on the 'prognostic politics' of environmental futures, Mathews and Barnes (2016) argue that practices of prediction or prognostication "always encounter limitations, making each future form potentially unstable, open to remaking, unmaking, or reinterpretation" (p. 21).

References

Adams, V., Murphy, M. and Clarke, A.E. 2009. Anticipation: Technoscience, life, affect, temporality. *Subjectivity*. 28(1), pp. 246–265.

Alley, K.D., Hile, R. and Mitra, C. 2014. Visualizing hydropower across the Himalayas: Mapping in a time of regulatory decline. *HIMALAYA, the Journal of the Association for Nepal and Himalayan Studies*. 34(2), pp. 52–66.

Anderson, B. 2010. Preemption, precaution, preparedness: Anticipatory action and future geographies. *Progress in Human Geography*. 34(6), pp. 777–798.

Appadurai, A. 2011. The ghost in the financial machine. *Public Culture*. 23(3 65), pp. 517–539.

Barry, A. 2013. *Material politics: Disputes along the pipeline*. West Sussex: John Wiley & Sons.

Bear, L. 2015a. Capitalist divination: Popularist speculators and technologies of imagination on the Hooghly River. *Comparative Studies of South Asia, Africa and the Middle East*. 35(3), pp. 408–423.

Bear, L. 2015b. *Navigating austerity: Currents of debt along a South Asian river*. Stanford: Stanford University Press.

Bear, L., Birla, R. and Puri, S.S. 2015. Speculation: Futures and capitalism in India. *Comparative Studies of South Asia, Africa and the Middle East*. 35(3), pp. 387–391.

Bhushal, R. 2016. Dreams of hydropower dollars: Koshi Basin Part 3. *The Third Pole*, 1 July [Online]. Available from: www.thethirdpole.net/2016/07/01/dreams-of-hydropower-dollars-koshi-basin-part-3/ [Accessed 15 May 2017].

Bloch, E. 1986. *The principle of hope* (Vol. 3). Cambridge: MIT Press.

Butler, C.J. 2016. *Knowledge, nature, and nationalism: The Upper Karnali dam in Nepal*. PhD Dissertation. University of California, Santa Cruz.

Butler, C.J. and Rest, M. (2017). Calculating risk, denying uncertainty: Seismicity and hydropower development in Nepal. *HIMALAYA, the Journal of the Association for Nepal and Himalayan Studies*. 37(2) pp. 15–25.

Campbell, B. 2010. Rhetorical routes for development: A road project in Nepal. *Contemporary South Asia*. 18(3), pp. 267–279.

Carse, A. 2014. *Beyond the big ditch: Politics, ecology, and infrastructure at the Panama Canal*. Cambridge: MIT Press.

Carse, A. 2017. An infrastructural event: Making sense of Panama's drought. *Water Alternatives*. 10(3), pp. 888–909.

Chilime Hydropower Company Ltd. 2014. *Annual Reports/Press Releases*. Company Website [Online]. Available from: www.chilime.com.np [Accessed 15 November 2016].

Clark, N., 2011. *Inhuman nature: Sociable life on a dynamic planet*. London: Sage Publications.

Collier, S., Mizes, J.C. and von Schnitzler, A. 2016. Preface: Public infrastructures/infrastructural publics. *Limn*. 7(1) [Online]. Available from: http://limn.it/preface-public-infrastructures-infrastructural-publics/ [Accessed 15 May 2017].

Cross, J. 2014. *Dream zones: Anticipating capitalism and development in India*. London: Pluto Press.

Cross, J. 2015. The economy of anticipation: Hope, infrastructure, and economic zones in South India. *Comparative Studies of South Asia, Africa and the Middle East*. 35(3), pp. 424–437.

Dixit, A. and Gyawali, D. 2010. Nepal's constructive dialogue on dams and development. *Water Alternatives.* 3(2), pp. 106–123.

Drew, G. 2017. *River dialogues: Hindu faith and the political ecology of dams on the sacred Ganga.* Tucson: University of Arizona Press.

Folch, C. 2016. The nature of sovereignty in the Anthropocene: Hydroelectric lessons of struggle, otherness, and economics from Paraguay. *Current Anthropology.* 57(5), pp. 565–585.

Forbes, A.A. 1999. The importance of being local: Villagers, NGOs, and the World Bank in the Arun Valley, Nepal. *Identities Global Studies in Culture and Power.* 6(2–3), pp. 319–344.

Gergan, M. D. 2016. *Precarity and possibility at the margins: Hazards, infrastructure, and indigenous politics in Sikkim, India.* Doctoral Dissertation, The University of North Carolina at Chapel Hill.

Golub, A. 2014. *Leviathans at the gold mine: Creating indigenous and corporate actors in Papua New Guinea.* Durham: Duke University Press.

Government of Nepal. 2016. *National energy crisis reduction and electricity development decade plan.* Kathmandu: Government of Nepal.

Government of Nepal_National Planning Commission. 2015. *Post disaster needs assessment.* Kathmandu: Government of Nepal.

Gupta, A. 2015. Suspension. *Cultural Anthropology: Theorizing the Contemporary,* 24 September [Online]. Available from: https://culanth.org/fieldsights/722-suspension [Accessed 15 November 2016].

Guyer, J. 2007. Prophecy and the near future: Thoughts on macroeconomic, evangelical, and punctuated time. *American Ethnologist.* 34(3), pp. 409–421.

Guyer, J. 2017. When and how does hope spring eternal in personal and popular economics? thoughts from Weste Africa to America. In: Miyazaki, H. and Swedberg, R., eds. *The economy of hope.* Philadelphia: University of Pennsylvania Press, pp. 147–171.

Gyawali, D. 2003. *Rivers, technology and society.* Kathmandu: Himal Books.

Gyawali, D. 2006. Hype and hydro (and, at last, some hope) in the Himalaya. In: Verweij, M. and Thompson, M., eds. *Clumsy solutions for a complex world: Governance, politics and plural perceptions.* London: Palgrave McMillan, pp. 61–85.

Harvey, P. and Knox, H. 2012. The enchantments of infrastructure. *Mobilities.* 7(4), pp. 521–536.

Hetherington, K. 2016. Surveying the future perfect: Anthropology, development and the promise of infrastructure. In: Harvey, P., Jensen, C.B. and Morita, A., eds. *Infrastructures and social complexity: A companion.* London: Routledge, pp. 40–50.

Hilton, I. 2015. Nepal's dangerous dams. *The New Yorker,* 30 April [Online]. Available from: www.newyorker.com/news/news-desk/nepals-dangerous-dams [Accessed 1 May 2016].

Howe, C., Lockrem, J., Appel, H., Hackett, E., Boyer, D., Hall, R. and Ballestero, A. 2016. Paradoxical infrastructures: Ruins, retrofit, and risk. *Science, Technology, & Human Values.* 41(3), pp. 547–565.

Huber, A., Gorostiza, S., Kotsila, P., Beltrán, M.J. and Armiero, M. 2017. Beyond 'socially constructed' disasters: Re-politicizing the debate on large dams through a political ecology of risk. *Capitalism Nature Socialism.* 28(3), pp. 48–68.

Huber, A. and Joshi, D. 2015. Hydropower, anti-politics, and the opening of new political spaces in the eastern Himalayas. *World Development.* 76, pp. 13–25.

Immerzeel, W.W., Van Beek, L.P. and Bierkens, M.F. 2010. Climate change will affect the Asian water towers. *Science.* 328(5984), pp. 1382–1385.

Kargel, J.S., Leonard, G.J., Shugar, D.H., Haritashya, U.K., Bevington, A., Fielding, E.J., Fujita, K., Geertsema, M., Miles, E.S., Steiner, J. and Anderson, E. 2016. Geomorphic and geologic controls of geohazards induced by Nepal's 2015 Gorkha earthquake. *Science.* 351(6269).

The Kathmandu Post. 2015. Energy sector suffered losses of Rs18.75b due to quake. *The Kathmandu Post,* 10 June: 5.

The Kathmandu Post. 2016. Power crisis to end by government lights: Energy emergency. *The Kathmandu Post*, 19 February: 1.

Klein, N. 2007. *The shock doctrine: The rise of disaster capitalism.* New York: Macmillan.

Knox, H.C. 2017. Affective infrastructures and the political imagination. *Public Culture.* 29(2), pp. 363–384.

Lakoff, A. 2010. Our energy production system: too big to fail. *Huffington Post*, 22 May 2010, [Online]. Available from: https://www.huffingtonpost.com/andrew-lakoff/our-energy-production-sys_b_586119.html [Accessed 1 February 2017].

Larkin, B. 2013. The politics and poetics of infrastructure. *Annual Review of Anthropology.* 42, pp. 327–343.

Lord, A. 2014. Making a 'hydropower nation': Subjectivity, mobility, and work in the Nepalese hydroscape. *HIMALAYA, the Journal of the Association for Nepal and Himalayan Studies.* 34(2), pp. 111–121.

Lord, A. 2016. Citizens of a hydropower nation: Territory and agency at the frontiers of hydropower development in Nepal. *Economic Anthropology.* 3(1), pp. 145–160.

Lord, A. 2017. Humility and hubris in hydropower. *Limn*, 9(1) [Online]. Available from: https://limn.it/humility-and-hubris-in-hydropower/ [Accessed 15 Jan 2017].

Mathews, A.S. and Barnes, J. 2016. Prognosis: Visions of environmental futures. *Journal of the Royal Anthropological Institute.* 22(S1), pp. 9–26.

McGoey, L. 2012. The logic of strategic ignorance. *The British Journal of Sociology.* 63(3), pp. 533–576.

Menga, F. 2015. Building a nation through a dam: The case of Rogun in Tajikistan. *Nationalities Papers.* 43(3), pp. 479–494.

Miyazaki, H. 2004. *The method of hope: Anthropology, philosophy, and Fijian knowledge.* Stanford: Stanford University Press.

Miyazaki, H. 2006. Economy of dreams: Hope in global capitalism and its critiques. *Cultural Anthropology.* 21(2), pp. 147–172.

Miyazaki, H. 2017. The economy of hope: An introduction. In: Miyazaki, H. and Swedberg, R., eds. *The economy of hope.* Philadelphia: University of Pennsylvania Press, pp. 1–36.

Molle, F., Mollinga, P. and Wester, P. 2009. Hydraulic bureaucracies and the Hydraulic Mission: Flows of Water, Flows of Power. *Water Alternatives.* 2(3), pp. 328–349.

Nixon, R. 2010. Unimagined communities: Developmental refugees, megadams and monumental modernity. *New Formations.* 69(69), pp. 62–80.

Ong, A. 2006. *Neoliberalism as exception: Mutations in citizenship and sovereignty.* Durham: Duke University Press.

Pandey, B. 1996. Local benefits from hydro development. *Studies in Nepali History and Society.* 1(2), pp. 313–344.

Pandey, S. 2016. The energy emergency. *República*, 29 March: 1.

Pangeni, R. 2015. Fuel crisis hits infrastructure projects. *República*, 8 October: 2.

Pattison, P. 2015. In Nepal, $1bn impact of strikes over constitution 'worse than earthquakes'. *The Guardian*, 5 October [Online]. Available from: www.theguardian.com/global-development/2015/oct/05/nepal-1bn-dollars-impact-economy-strikes-over-constitution-worse-than-earthquakes [Accessed 15 November 2015].

Pigg, S.L. 1992. Inventing social categories through place: Social representations and development in Nepal. *Comparative Studies in Society and History.* 34(3), pp. 491–513.

Rajak, D. 2011. *In good company: An anatomy of corporate social responsibility.* Stanford: Stanford University Press.

Reeves, M. 2016. Infrastructural hope: Anticipating 'independent roads' and territorial integrity in southern Kyrgyzstan. *Ethnos.* 82(4), pp. 711–737.

Republica. 2015. IPO draws overwhelming responses from investors. *República*, 5 March: 4.

Rest, M. 2012. Generating power: Debates on development around the Nepalese Arun-3 hydropower project. *Contemporary South Asia.* 20(1), pp. 105–111.

Rest, M., Lord, A. and Butler C. 2015. The damage done and the dams to come. *Cultural Anthropology: Hot Spots,* 14 October [Online]. Available from: https://culanth.org/fieldsights/730-the-damage-done-and-the-dams-to-come [Accessed 15 November 2016].

Sangraula, B. 2015. Nepal accuses India of an economic blockade as border trade freezes up. *Christian Science Monitor,* 28 September [Online]. Available from: www.csmonitor.com/World/Asia-South-Central/2015/0928/Nepal-accuses-India-of-an-economic-blockade-as-border-trade-freezes-up [Accessed 15 November 2016].

Shamir, R. 2008. The age of responsibilization: On market-embedded morality. *Economy and Society.* 37(1), pp. 1–19.

ShareSansar. 2014. Interview with Kul Man Ghising: Part III of the Chilime Series. *ShareSansar Website,* 9 June [Online]. Available from: www.sharesansar.com/c/part-iii-of-the-chilime-series.html [Accessed 10 July 2016].

ShareSansar. 2015. HIDCL IPO worth NRs 2 Arba from Kartik 12. *ShareSansar Website,* 6 November [Online]. Available from: www.sharesansar.com/ [Accessed 10 July 2016].

ShareSansar. 2016. IPOs worth 603 million in the pipeline. *ShareSansar Website,* 5 September [Online]. Available from: www.sharesansar.com/ [Accessed 15 September 2016].

ShareSansar. 2017. Listed Companies: Hydropower. *ShareSansar Website* [Online]. Available from: www.sharesansar.com/company-list/ [Accessed 15 September 2016].

Shrestha, B. 2015. Upper Tamakoshi damage manageable. *Spotlight Nepal,* 11 September [Online]. Available from: www.spotlightnepal.com/2015/09/11/upper-tamakoshi-damage-manageable/ [Accessed 15 November 2016].

Shrestha, B., Lord, A., Mukherji, A., Shrestha, R., Yadav, L. and Rai, N. 2016. *Benefit sharing and sustainable hydropower: Lessons from Nepal.* Kathmandu: The International Center for Integrated Mountain Development.

Simpson, E. 2013. *The political biography of an earthquake: Aftermath and amnesia in Gujarat, India.* London: Hurst.

Sneath, D., Holbraad, M. and Pedersen, M.A. 2009. Technologies of the imagination: An introduction. *Ethnos.* 74(1), pp. 5–30.

Swyngedouw, E. 2015. *Liquid power: Contested hydro-modernities in twentieth-century Spain.* Cambridge: MIT Press.

Tamot, S. 2014. A bug's life. *The Kathmandu Post,* 29 December: 7.

Thapa, M.G. and Shrestha, K.S. 2015. Natural disasters and Nepal's energy security. *New York Times,* 25 May 2015, DotEarth Blog [Online]. Available from: https://dotearth.blogs.nytimes.com/2015/05/25/one-two-punch-of-earthquakes-and-landslides-exposes-hydropower-vulnerability-in-nepal/ [Accessed 15 October 2016].

Thompson, M., Warburton, M. and Hatley, T. 2007 [1986]. *Uncertainty on a Himalayan scale.* Kathmandu: Himal Books.

Tsing, A.L. 2005. *Friction: An ethnography of global connection.* Princeton: Princeton University Press.

Welker, M. 2014. *Enacting the corporation: An american mining firm in post-authoritarian Indonesia.* Berkeley: University of California Press.

Welker, M. and Wood, D. 2011. Shareholder activism and alienation. *Current Anthropology.* 52(S3), pp. S57–S69.

Weszkalnys, G. 2016. A doubtful hope: Resource affect in a future oil economy. *Journal of the Royal Anthropological Institute.* 22(S1), pp. 127–146.

White, R. 1995. *The organic machine: The remaking of the Columbia River.* New York: Macmillan.

Zarfl, C., Lumsdon, A.E., Berlekamp, J., Tydecks, L. and Tockner, K. 2015. A global boom in hydropower dam construction. *Aquatic Sciences.* 77(1), pp. 161–170.

12 Irrigational illusions, national delusions and idealised constructions of water, agriculture and society in Southeast Asia

The case of Thailand

David J.H. Blake[1]

Introduction

The entire Mekong region[2] of Southeast Asia is currently undergoing an unprecedented boom in hydraulic infrastructure development, including both hydropower and irrigation components. The material and discursive roots of the latest phase of a regionally distributed "hydraulic mission"[3] (Mirumachi, 2012; Molle et al., 2009b) can be partially traced back to several key moments in the mid to late twentieth century, which were critical in laying the conditions for the post-2005 reinvigorated pursuit of hydraulic development presently underway (Blake and Robins, 2016). The sheer scale and ambition of the numerous hydraulic construction projects, both planned and under construction, are quite remarkable in their complexity and transformative implications for society and the environment. Much of the new development in the lowland areas is focused on irrigation development, yet this sector receives proportionately much less critical attention from regional analysts of the water resources development paradigm compared to hydropower. Moreover, irrigation is frequently perceived by observers to represent a semi-benign form of hydraulic development, inherently assumed to benefit farmers as a privileged category of water users, stimulate the economy and contribute to solving problems of rural poverty, local water scarcity and increasingly, address climate change–induced drought (e.g. Mukherji et al., 2009). Hydropower, conversely, is unable to invoke the same benign image and has become a keenly contested hydraulic development intervention regionally, drawing considerable criticism from a wide and growing range of detractors, raising fears of a "race to the bottom" in the case of the Mekong basin (Matthews and Geheb, 2015; Hirsch, 2014).

Irrigation is a core component of nearly all of the individual Mekong nation's regional development strategies and policies (Johnston et al., 2010; Hoanh et al., 2009), with only Myanmar recently expressing a degree of reservation about irrigation's potential to solve rural problems and effectively develop the country. In a remarkable departure from the norm, national news reports suggested that the

190   *David J.H. Blake*

government was not planning to build any new irrigation schemes during its tenure (Htet Naing Zaw, 2016), and was going to cut the national budget for irrigation expenditure in 2017–18 by over 50%, signalling a shift in policy away from building new dams to focusing on the repair, operation and maintenance of existing projects (Htoo Thant, 2016). By marked contrast, in other Lower Mekong Basin (LMB) states there seems to be an insatiable thirst for increasing irrigation coverage, with a planned expansion of 1.6 million hectares over the next 20 years from 2010, with Thailand poised to add the greatest command area of over 900,000 ha that represents a greater area than the combined total of the other three countries' planned additional expansion (Hall et al., 2014). This contribution seeks to shed some light on diverse, non-conventional socio-political and cultural drivers of the modern irrigation development paradigm, suggesting that earlier explorations of the topic have not sufficiently teased out some of the more ideological and culturally embedded factors related to nationalism and how they may offer some explanation for prevailing social power relations.

Most of Thailand's proposed future irrigation expansion is focused within the Northeast region, which has the lowest irrigation provision relative to total arable land of all regions and is often portrayed as the most water scarce and impoverished. The incumbent military regime (2014 to present) under Prime Minister Prayuth Chan-ocha has repeatedly signalled its intentions to increase expenditure on the irrigation sector nationwide, in particular through the construction of a number of diversion projects, extracting water from transboundary rivers situated along its borders. According to a map published in the *Bangkok Post* using data from the Royal Irrigation Department (RID), Thailand plans to divert 51,560 million cubic metres of water via 13 large and 19 smaller pumping schemes including from the Salween and Mekong Rivers, which will supposedly benefit 3.43 million people nationwide and irrigate 9.8 million hectares of land (Wipatayotin and Wangkiat, 2016). The total costs of these diversion projects are estimated at 2,700 billion baht (approximately $84 billion) (Wangkiat, 2016), with the Mekong diversion alone, known nationally as the Khong-Loei-Chi-Mun project, estimated to boost the annual income of beneficiary households by about $1,950 each. The announcement of such ambitious domestic schemes in Thailand, often labelled "mega-projects", follows a well-trodden tradition of other similar schemes announced by various regimes over past decades (Blake, 2016; Molle and Floch, 2008). Most of these schemes have never reached fruition as the political cliques behind them have fallen from power first, although they are usually implemented in part, leading to considerable internal conflict from associated social and environmental impacts (Floch and Blake, 2011; Molle et al., 2009a; Floch et al., 2007).

The most controversial of the partly implemented mega-projects to date has been the Khong-Chi-Mun (KCM) project and its blueprint successor, the Water Grid project (Molle and Floch, 2008), originally proposed under the Thaksin Shinawatra (2001–06) led regime and reignited later under various proxy prime ministers, including during his sister Yingluk Shinawatra's term of office (2011– 14) (Blake, 2016). Both these projects were partly implemented but gradually became embroiled in widespread controversy for their socio-ecological impacts, as well as allegations of corruption, lack of transparency, low participation and other questionable governance indicators. As Molle (2005: 3) noted over a decade ago,

> [N]evertheless, while it would seem that Thailand has largely developed its water resources and the 'hydraulic mission' is coming to an end, further ambitious developments seem to be on the way (despite the alleged priority to demand-management announced in the Ninth National Plan [2002–2006]).

Given the present stage of Thailand's overall development paradigm, ongoing industrialisation and growth of the service sector, the concomitant growth of urban populations, a diminishing share of the agriculture sector in the national Gross Domestic Product (GDP) and continuous decline in farmer numbers, it seems somewhat paradoxical that the government should be seeking to greatly expand its irrigated area at this juncture, raising questions about the underlying motivations and logic of the government's declared water resources policies and strategy.

By providing some empirical observations to accompany a novel conceptual term, this chapter helps to elucidate the underpinnings and persistence of the hydraulic mission in Thailand, a mission which has steadily moved its focus from the national centre (i.e. Bangkok) to the periphery and is now actively spreading its influence to the territory of neighbouring states through comprehensive involvement in hydropower projects, transboundary basin water transfers and diversions[4] (see Matthews, 2012). At a more general level, this chapter contributes to recent literature conceptualising hydraulic development processes at a sub-national or transboundary level, in which nationalisms and issues related to formations of a national identity are conceived (e.g. Menga, 2015; Allouche, 2005). Menga (2016) has noted the paucity of research regarding how the formation of a national identity may overlap with large-scale hydraulic infrastructure construction. To this end, this study contributes to what Mollinga (2008) refers to as a political sociology of water resources management, where water control is understood to be at the heart of water resources management, conceived as a process of politically contested resource use. The case study is primarily situated in the domain classified as "the politics of water policy in the context of sovereign states" (Mollinga, 2008: 12).

Reviewing the literature, it seems that perspectives vary considerably in identifying the main drivers of the irrigation development paradigm in Thailand and its neighbours. Conventional analyses suggest that irrigation infrastructure expansion in Asia is mainly driven by instrumental factors such as population growth, urbanisation, globalisation, agricultural diversification, changing diets and climate change (Faures and Mukherji, 2009). Barker and Molle (2004) argued that during a period of rapid growth in Asian irrigation between the 1960s and 80s, prioritising national food security was a major driver, but after a slowdown in sectoral investment growth during the 1990s, rural livelihoods and sustainability concerns became the overriding drivers as water resources scarcity was recognised and the national focus switched from supply-led to demand-oriented development. However, this broad brush continent-level analysis does not seem to be borne out by the situation observed in Thailand, or its neighbours, over the last few decades, where state infrastructural investment has increased and the main bureaucracy involved is still obsessed with irrigated area expansion as an end in itself, rather than a means to an end.

Taking a different analytical approach, researchers using frameworks within political ecology have examined the dominant discourse of irrigation development to suggest that powerful state actors tend to adopt a number of "meta-justifications" in pursuit of expanding irrigation command areas, diverting water across basins and national boundaries and generally enhancing hydraulic control (Molle et al., 2009a; Molle and Floch, 2008; Molle, 2007). These meta-justifications include strengthening national security, "modernization", rural poverty alleviation, food and water security, self-sufficiency and export substitution concerns as some of the most significant discursive drivers of pharaonic new hydraulic projects. More specifically, Molle (2008: 217) has identified "ideology and state building" as a key driver of hydraulic development, leading to the "overbuilding" of river basins, a factor which this paper seeks to elaborate. Yet another common justification used by state actors, namely symbolic support from the monarchy, which for the case of Thailand was personified through the late Rama IX,[5] is frequently invoked by state actors pushing new schemes; this aspect has been covered in more detail elsewhere (Blake, 2017; Blake, 2015b; Blake, 2012) and will only be touched upon in this paper. To better understand the present context, a brief historical examination of the irrigation development paradigm in Thailand helps to ground some of the actors and emergence of key institutions that have shaped the field.

Dam building and irrigation – so obviously a good thing for Thai society?

Contemporary large-scale irrigation approaches in Thailand partly have their origins in attempts to comprehensively control and transform the water landscape of the centrally located Chao Phraya delta during the latter years of the reign of Rama V, or King Chulalongkorn, during the first decade of the 20th century (Brummelhuis, 2005; Kaida, 1978), relying on technical expertise from Dutch and British engineers, amongst other European nations involved as Siam government advisors. Irrigation rose to prominence as a major socio-economic and technological intervention in rural development strategies during the post–World War Two period of regional colonial transition. It was precipitated, in large measure, by the United States of America's geo-political and ideological interests in the region, later to become a full-blown armed struggle against the Soviet Union's influence and suppression of communism more generally (Barker and Molle, 2004). Thailand has enthusiastically constructed large dams and public irrigation schemes in all regions since the late 1950s, beginning with the iconic Bhumibol Dam on the northern Ping River for hydropower, closely followed by the long-planned Chainat Dam and Greater Chao Phraya Irrigation Project, which according to Molle (2003), heralded a so-called "irrigation phase" in the Central Plains' development paradigm.

The focus on large-scale irrigation project construction gradually spread from the hydraulic "core" (i.e. Bangkok and the Chao Phraya delta) to the periphery of the Thai kingdom from the 1960s on, as Thailand became a favoured client state for mainstream Western developmentalism. Most irrigation and hydropower dams were built with the financial and technical assistance of Western nations, both as bilateral and multilateral aid packages, invariably described by proponents as rural development and poverty alleviation projects. The "arid and poor" Northeast region, separated from the Central Plains by a range of mountains to the west and bounded by the Mekong River to the north and east, garnered some of the closest attention by external agencies due to perceived risks of communist insurgency, but also posed some of the biggest obstacles with regards to implementing irrigation on a large scale, in part due to its topography, geology and hydrology. The US government, acting through a range of domestic agencies and under the auspices of the United Nations Economic Committee for Asia and the Far East (UNECAFE), helped establish several regional and national-level institutions with interests in water resources management, including the Mekong Committee in 1957 (Chomchai, 1994). Establishing a complex techno-political network of actors, this institution became the vehicle through which an elaborate (but poorly quantified in terms of costs and benefits) scheme to build a cascade of large-scale dams along the Mekong mainstream (and numerous tributaries besides) was planned, that became internationally known as the Mekong Project (Sneddon, 2015). This vastly ambitious scheme from the Cold War era, to a considerable extent, still shapes the visions and imagination of national planners and strategists in transforming the region into an irrigated verdant utopia (Blake, 2012; Molle et al., 2009a; Guyot, 1987).

Given the apparent maturity of the Thai hydraulic infrastructure development paradigm over the course of the last century combined with a chronic underperformance technically and economically of existing irrigation schemes when compared against other infrastructural investments, studies indicate that irrigation has delivered the smallest relative impact on poverty reduction and productivity growth in agriculture (Fan et al., 2004; Grachangnetara and Bumrungtham, 2003). Coupled with a low relative contribution to national GDP by the agricultural sector (consistently varying between 10% and 12% since 1993) and ongoing transition of the rural population away from predominantly agriculture-based livelihoods (Floch and Molle, 2013; Rigg, 2005), one could legitimately ponder why the state should still seek to invest so much public capital in the irrigation development sector, especially in marginal areas like the Northeast where performance has traditionally proven to be the lowest. It has been shown that the state has attempted every possible hydraulic engineering combination and permutation for developing agricultural water supply, and implemented these technologies at a range of scales over the last six decades (Molle et al.,

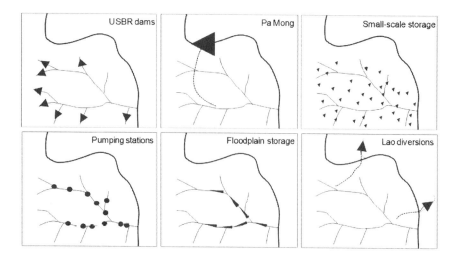

Figure 12.1 Representation of a range of hydraulic infrastructure development options proposed, planned and/or implemented at different time periods in the Northeast
(Source: Molle et al. 2009a, Figure 10.5: 273)

2009a, see Fig, 12.1), yet all the interventions have essentially demonstrated the same lacklustre technological performance and disappointing socio-economic returns (Floch and Molle, 2013). Berkoff (2007: 189), in trying to account for the consistently over-optimistic economic forecasts associated with World Bank irrigation projects in Thailand and elsewhere, points out that "institutional incentives for going ahead with a project often outweigh any doubts associated with the economic analysis" and that political dynamics ("irrigation is so obviously a good thing – who can be against it?") and beneficiary farmer self-interest to proceed with development explain the rest.

Introducing the concept of "irrigationalism"

In trying to explain one of the key drivers of the modern irrigation development paradigm, Blake (2012) hypothesises that contemporary Thailand exhibits many of the hallmarks of what he terms an irrigational society i.e. one that inherently professes an ideology of "irrigationalism", which can be defined as

> a dominant ideology of irrigation developmentalism that incorporates utopian notions of agricultural modernism, technocentrism and more primordial nationalistic sets of beliefs, values and actions in [Thai] society about the potential of irrigation; performing important transformative functions in an elite-oriented project of domination over nature and society.
>
> (ibid.: 99)

State-sanctioned and executed irrigation development projects embody distinctive elements of mystification, illusion and falsity in their outward dissemination from the centre to the periphery, forming an integral part of what Anderson (2006) refers to as "the imagined community" of Thai nationhood. This includes both strands of elite calls for embracing technological modernity (i.e. agricultural high modernism in James Scott's terminology) and simultaneously calling upon supposed traditional Thai superiority in agricultural water management techniques that compete with neighbouring states' claims (i.e. a form of hydraulic primordialism).

The ideology of irrigationalism is argued to have been deliberately and systematically cultivated by Thai state-allied elites over several generations, buttressing irrigation as a "technology of control", thus providing a partial explanation as to why the main hydraulic bureaucracies have engaged in a ceaseless planning and implementation cycle of developing irrigation projects at all scales, removing the need for a sound socio-economic rationale or burden of proof of genuine end-user demand. This term builds upon earlier work that suggested an ideological basis for irrigation development, most particularly that of Adams (1991), who coined the related term "irrigationism" as a synonym for ideologically motivated and external expert–mediated blueprint-type irrigation development found in certain Sub-Saharan African countries and, in a different context, Hamilton McKenzie (2009), who proposed the term "irrigationist philosophy" to refer to a zealous drive for irrigated agricultural development in late 19th-century California and Australia's Murray-Darling Basin, which had adopted the same founding principles. These developments were based on firstly, an independent and civilised yeoman farmer ideal; secondly, the superiority of a technologically progressive and scientifically inspired irrigated agriculture, over that of broad-acre farming; and thirdly, fulfilling a preternatural mission to transform wild, barren, dryland landscapes into well-ordered, green and productive land tracts. The chief evangelisers of this civilising agricultural mission were a pair of Canadian brothers (the Chaffeys), who developed a prescriptive "Red Book" in which their vision for irrigation settlements was laid out in a context where "the irrigationist movement was given a tradition, a sense of historical trajectory, which caught the readers in a sense of destiny" (Hamilton-McKenzie, 2009: 35), not unlike the Zionist ideology calling for increased water and land control in modern-day Israel and the West Bank territories (see Lipchin, 2007; Kartin, 2001).

Predating these conceptual understandings of how irrigation development and water resources development can be underpinned by state-mediated ideology was the work of Aaron Wiener, who proposed a dichotomous distinction between ideological and pragmatic approaches to problem solving, claiming the former approach was derived from "the systematic application of a priori type of principles to an intentionally highly simplified planning space" (Wiener, 1972: 72). He stresses how these underlying principles "often have their roots in broader national, religious, moral or political tenets that have been taken over uncritically from an irrelevant past" (ibid.: 26). The antithesis of the ideological

196   *David J.H. Blake*

approach is the pragmatic approach, he argues, where water resources planners seek an unbiased definition of the planning space, its problems, dimensions, structure, trends and constraints, before deciding on the type of intervention model to select that both best represents the functional relationship of the space and its responses to change.

The degree to which the Thai state and its allied institutions have adopted an ideological approach over a more pragmatic approach to developing and managing its agricultural water resources is discussed in the following section.

Hydraulic agriculture development, monarchy and "Thainess"

> The basic impetus for this drive is the same one that has been responsible for progress in the past: the sense of 'Thainess' engendered by common origins, culture and aspirations. The manifestations of that spirit can be seen in the fervour with which many of the problems which lie in the way of Thailand's total development are being remedied in a rational manner according to objectives laid out in government plans.
> (Source: Office of the Prime Minister, 1979: 158)

The above quote extracted from a richly illustrated book titled *Thailand into the 80s*, a publication that was no doubt produced as a blunt propaganda tool to market the merits of Thailand to the outside world as an investment-friendly destination to transnational capital at the outset of a prolonged economic boom period, hints at the supposedly rational approach underpinning the cultural concerns informing the government's "total development" plans. It introduces the concept of "Thainess" as a much-vaunted but rather pliable term, generally used in an overwhelmingly positive sense to describe the essence, characteristics, ideology and values of a uniquely Thai national identity (see Streckfuss, 2011; Sattayanurak, 2005; Reynolds, 2002). The notion that ethnic Thai ancestors were historical masters of water resources management and control to rival the Angkorian empire has often been suggested by various elite arbiters of Thainess as a core part of an immutable national identity, an idea that has been reified on occasions by external organisations such as in the case of an Asian Development Bank–hosted website which celebrated Thailand's "regional leadership" in the water sector "after centuries of involvement with the issue" (Whaley, 2005).

As Samudavanija (2002: 60) emphasises, echoing Anderson's (2006) and others' interpretation, "the official versions of national identity and culture were constructs based on the creation of an historical imaginaire." The post–World War Two Thai state, as an authoritarian polity that veered towards the despotic under the contiguous regimes of one strongman military general after another (Chaloemtiarana, 2007), embraced functions of depoliticisation, bureaucratisation and socialisation of a new nation with gusto, processes which have been inextricably linked with the state ever since. One of the key components in the elite construction of a coherent national identity for the modern Thai state was to call upon imagined pasts, including popular myths that Thais are naturally gifted water resource managers, as professed through an innate superiority in communally managed irrigated rice cultivation and evidenced by preserved documentation supposedly dating back over eight centuries allowing the evolution of complex management rules, outlined below (Surarerks, 2006). This could be interpreted as an elite attempt to instil a source of national pride in present agricultural water management skills, traditions and cultural values associated with guidance from a wise, just and paternalistic monarchical institution.

One interpretation of why Thai state elites might conspire to convert extensive areas of land in the Northeast presently classified as "rainfed" into bureaucratically managed irrigation schemes could be derived from Scott's (1998) concept of authoritarian high modernism. High modernism, Scott contends, is premised on a strong elite belief in the benefits of scientific and technical progress associated with industrialisation in Western Europe and North America, in this case as it could be applied to agricultural intensification of production.

> At its center was a supreme self-confidence about continued linear progress, the development of scientific and technical knowledge, the expansion of production, the rational design of social order, the growing satisfaction of human needs, and, not least, an increasing control over nature (including human nature) commensurate with scientific understanding of natural laws.
> (Scott, 1998: 89–90)

As Scott was at pains to point out, agricultural high modernism was regarded as a faith tenet across a wide range of political ideologies that not only promised far-reaching economic benefits, but just as importantly for the state, offered opportunities for increased tax revenue, alongside radical simplification and improved legibility of the periphery.

However, as useful as Scott's concept of authoritarian high modernism may be in accounting for a state-centric paradigm based upon grandiose hydraulic infrastructure development in Thailand, such an explanation only goes so far. It fails to fully account for equally persistent clarion calls by elites to pursue such projects as a means to resurrect or rediscover some vaguely defined golden past, such as that invoked through the Suwannaphum legend (Keyes, 1977) and various popular or nationalistic tales concerning long-deceased powerful monarchs. This is evident in the narratives constructed around King Mengrai and the "democratic" nature of "muang fai" local irrigation technology of Northern Thailand, that occasionally invoke in the present at opportune moments.[6] King Mengrai reputedly ruled the Lanna Kingdom of present-day northern Thailand in the late 13th century and is credited with codifying irrigation management rules and laws (called "Mengrai Sart"), that are considered by some Thai academics as a national historical heirloom to future generations, providing a blueprint for modern irrigation management practices (e.g. Surarerks, 2006). Indeed, the so-called People's Irrigation Act B.E. 2482 (1939) that nominally allowed for local community-managed irrigation systems and is still occasionally referred to by the RID would seem to acknowledge a historical connection stretching back to Mengrai Sart in its formulation (Ounvichit, 2005), despite an

198 David J.H. Blake

evident lack of genuine farmer participation in modern irrigation management above the level of tertiary canals or small-scale gravity-fed irrigation projects or proliferation of bureaucratically sanctioned and merely perfunctory water user's groups (Blake, 2012; Neef, 2008; Schluter, 2006).

As modernity and development have challenged notions of social hierarchy and traditional elite rule in Thailand, so have efforts to rewrite history and incorporate interpretations of the past into present social relations. For example, Rigg and Ritchie (2002) demonstrate how elite constructions of a popularly imagined rural past suffuse ideas and practices in the present, which have in turn influenced the private sector, civil society, parastatal and government actors, through development narratives and strategies alike. As the above authors contend, "From their earliest years in school, when children are taught that farmers are the 'backbone' of the nation, the special character of farmers, farming and rural areas is stressed" (ibid.: 363). Tracing constructions of Thai rurality back to the 13th century, they cite the historically controversial Ramkhamhaeng inscription[7] as a major influence and argue that modern expressions of agricultural production and rurality have frequently been co-opted ("from within and without") for purposes of consumption and nation building. Indeed, the stone inscription has become a "key building block" and "central column in the edifice of Thai nationhood" (Rigg and Ritchie, 2002: 361), in particular the oft-quoted lines, "In the time of King Ramkhamhaeng, this land of Sukhothai is thriving. In the water there is fish, in the fields there is rice." As Blake (2012) has noted, this apocryphal passage is routinely cited by a variety of actors in Thai public discourse, appearing in official government publications, academic books, political speeches, TV shows, radio programmes and other media forms.

The building of nations around idealised notions of an agriculturally based society under the rule of a benevolent king has ancient precedents elsewhere in Asia, in particular China where the early agrarian philosophy of "agriculturalism" arose. Closely associated with peasant utopian communitarianism and egalitarianism, the ideology of agriculturalism has also been termed the School of Agrarianism (Deutsch and Bontekoe, 1997). According to Sellman (2002: 76), "the means by which the early sage kings led their people was to put agriculture before all other affairs. . . . the reason why Hou Ji undertook agriculture was because he understood it to be the root of instructing the masses." Agricultural development was believed to be the key to a stable and prosperous society, with agriculturalism being viewed as an essence of the Chinese identity, maintains Gladney (2004). It was based upon a notion of "people's natural propensity to farm" and resonates with associations of irrigated agriculture with higher civilisation and instilling a moral ideal of farming, as emphasised in examples offered by Molle et al. (2009b: 331).

I would argue that a coherent modern parallel to ancient China can be identified under the "hidden hand" guidance of King Bhumibol's simultaneously symbolic, practical and paternalistic rule, in which the Thai state has demonstrated a consistent bias towards advocating a preservation of or even return to a form of agriculturalism, in which irrigational tendencies are considered a vital sub-component. The elite's fascination with primordial constructions of an imaginary agriculture-based nation is tangibly expressed in many contemporary state programmes, strategies and policies, perhaps the most profound and far-reaching of which has been the royal-backed philosophy of a "Sufficiency Economy",[8] which was prominently incorporated into at least three National Economic and Social Development Plans (Intravisit, 2005) and runs parallel with promoting a hubristic irrigation development paradigm. Also indispensable has been the hegemonic "Royally-Initiated Projects" (Chanida, 2007), which have implemented largely inscrutable irrigation projects in their thousands, often in close cooperation with the RID (Blake, 2015). Further, I would posit that irrigational worldviews form an indispensable link with utopian agriculturalist visions, with implications towards steering collective national self-identity and stirring the public imagination.

Such a view is supported by arguments made by Fong that Thailand represents an example of a guardian state, "that protects itself from threats, but less from tangible military threats than ideological threats" (2009: 674), in particular perceived threats to the widely revered monarchical institution that sits atop the hierarchical edifice of Thai society, symbolised most tangibly in the ubiquitous form of the late, semi-deified King Bhumibol (Jackson, 2010). Like several other critical scholars of Thailand (e.g. Chaloemtiarana, 2007; Samudavanija, 2002), Fong traces the modern epoch of nation construction back to the despotic prime minister, Field Marshall Sarit Thanarat, during the late 1950s and early 60s. Sarit was seminal in evoking a series of primordial themes "to generate for the monarch and monarchy a rich, sacred and mystical link to a glorious, Buddhist and imperialistic Siamese past" (Fong, 2009: 675). Fong argues that the employment of diverse primordial simulacra are made visible to the people by the state, and effectively function as a ready supply of "dormant nostalgia" to be called upon to "amplify the merits of the nation". Examples of the primordial themes he identified include reviving a number of lapsed royal spectacles and ceremonies from the absolute monarchy era, as a means to aid military-bureaucratic elite efforts to sacralise the king and cement his persona as both a devaraja (a god-king in the Hindu-Brahman culture of kingship) and dhammaraja (a selfless king who rules according to the ancient Buddhist code of dhamma) in the minds of his subjects, including the Royal Ploughing Ceremony (phra rathcha pithi pheutcha mongkhon) in May. Embodying Buddhist and Brahmanical themes, the ceremony is designed to predict the forthcoming wet season rainfall and subsequent bounty of the crop harvest, with a history dating back to the Sukhothai kingdom (1238–1438) when subjects widely believed the sacral king could influence weather patterns, and thus the wellbeing of the nation. Presided over by King Bhumibol for decades, and latterly the Crown Prince, a pair of oxen plough a small patch of soil on Sanam Luang near the Grand Palace, while auspicious rice seed is ceremoniously scattered behind. The ceremony was reinstated as a major annual event in 1960 under Sarit's direct orders, with the king presiding over a lavish state-sponsored ritual, with officials and celebrants clothed in period attire and the proceedings broadcast to the nation by multiple Thai state-controlled media channels. Sarit was also responsible for reviving the practice of citizens crawling in front of royalty during audiences, a gesture of submission associated by Wittfogel (1981) with classic hydraulic despots.

Irrigation development, in so far as it is inherently assumed to improve economic, water and food security within Thailand's dominant discourse, has long been more conventionally securitised as a national security issue at the highest levels, thus closing many avenues of internal discussion. Starting from the early 1960s, Bhumibol worked ceaselessly to identify, design and oversee the construction of thousands of irrigation schemes across the nation, from the micro to the macro scale, earning himself the government-granted title, the "Father of Water Resources Management" in 1996 (Blake, 2015). In speeches to the nation, the king often stressed how the country could survive without electricity, but would soon starve without water for irrigation, flagging it as a sine qua non of nationhood. In another context, Callahan (2006: 186), using an analysis of contemporary China, emphasises how "'security' is not about defending us so much as 'tell[ing] us who we must be'" and proposes the notion that, like states, the nation is a production, with nation-building seen as a political performance. Callahan takes the example of "National Humiliation Day"[9] in China to explore the use of time and temporality in nation-building projects by state elites past and present, extending beyond the production of nationalism to incorporate a consumption element amongst the Chinese population, as part of a symbolic economy that helps cement identity.

Conclusions

Referring to contemporary expressions of large-scale irrigation development directed from the bureaucratic centre, this chapter contends that irrigationalism has long formed a dominant "sacred cow" ideology, one that is inseparable from nationalist and monarchist-related belief systems that are at once naturalising and universalising within mainstream dominant accounts of Thai nationhood. Despite six decades of relentless top-down hydraulic development, which has seen all regions subject to vast state-led investment in construction of irrigation schemes, there seems to be scant evidence that this strategy has had any discernible positive impact on poverty reduction or economic growth, defying the assumptions and expectations of many international development actors. Menga (2016: 717), in concluding remarks, enquired whether it is helpful to borrow from Scott when interpreting "the current boom in the dam building sector as a 21st century revamp of the ideology of high modernism". I would posit, on the basis of the Thai case, that one should be rather cautious before drawing this conclusion. I have argued that the specific ideological dimensions of the hydraulic mission have been largely overlooked by regional social science analysts, especially the degree to which water resources has been a determining factor, both concretely and discursively in cementing national elite goals of greater socio-political control. Conceptualising the persistent state-led irrigation expansion phenomenon as an expression of irrigationalism allows it to be understood as not merely an ideology of authoritarian high modernism, but

Irrigational illusions in Southeast Asia   201

also recognises the fact that it draws heavily on elite constructions of idealised pasts, incorporating mythologised ancient monarchs, naturalised social hierarchy and subordinate time-frozen peasant farmers portrayed as innate paragons of "Thainess" for practicing irrigated agriculture, thus safe-guarding Thai culture and demonstrating national unity in official representations. This has created a gnarly paradox for an "Upper Middle Income Country" currently under the rule of a military regime, proposing an economic transition to a post-agrarian and industrial state (termed "Thailand 4.0")[10] yet simultaneously employing outdated developmentalist narratives to justify a series of massive inter- and intra-basin water diversion schemes, primarily justified for irrigated agriculture and poverty eradication goals. Such hubristic pre-occupation with failed irrigational development that inherently crowds out alternative narratives from subaltern actors demonstrates the power of ideological approaches over more socially and environmentally pragmatic approaches.

Notes

1 Independent scholar. Correspondence address: 40 Woodrush Close, Taunton, Somerset, TA1 3XB United Kingdom. Email: djhblake@yahoo.co.uk.
2 The Mekong region is defined here as including the countries of Cambodia, the Lao People's Democratic Republic (PDR), Myanmar, Thailand, Vietnam, plus the two Chinese provinces of Guangxi and Yunnan, encompassing an area of 2.6 million square kilometers and home to over 320 million people.
3 The hydraulic mission has been concisely defined as "a phase of investment in dams, irrigation schemes and ground water extraction projects organized by the state to capture water" (Mirumachi, 2012).
4 It could be argued that Thailand is already well advanced in enacting a hydraulic mission in Laos, with several hydropower dams completed or under construction, which have been financed, designed, constructed and implemented by primarily Thai corporations, and a similar mission is in the early stages in Myanmar.
5 King Bhumibol Adulyadej, the ninth king in the Chakri dynasty, died on 13 October 2016 at the age of 88, after 70 years on the throne.
6 The spirit of King Mengrai was graphically invoked at the 2nd Asia-Pacific Water Summit held in Chiang Mai in May 2012, an event hosted by the Thai government. Plodprasop Suraswadi, the Deputy Prime Minister and chairman of the Water and Flood Management Commission, appeared in period costume playing the role of King Mengrai in a son et lumière extravaganza, in front of an audience comprised of hundreds of high-level foreign dignitaries and conference delegates. The entire event was fortuitously timed ahead of government efforts to raise investments funds for its controversial 350 billion baht national water development scheme (see www.apwf.org/weblog/public_html/weblog/information/archives/201305/000082.php and http://pattayatoday.net/news/thailand-news/plodprasop-to-portray-ex-king-at-water-summit/ Accessed 2 October 2016).
7 King Ramkhamhaeng was reputedly a great and benevolent monarch who ruled a kingdom located in present-day northern Thailand from 1279–98. His works and deeds are reputedly recorded on a stone inscription discovered in the 19th century, now located in the National Museum, which has sometimes been portrayed as the first major work of Thai literature (Chitkasem, 1999) and considered Thailand's "first constitution" in the mode of a Thai-style Magna Carta (Pramoj, 1990).
8 For critiques of the Sufficiency Economy philosophy developed by King Bhumibol, on both a practical and rhetorical level, refer to works by Walker (2010), Isager and Ivarsson (2010) and Intravisit (2005).

202 *David J.H. Blake*

9 National Humiliation Day is a public holiday in China first declared in 2001, that is more formally referred to as National Defense Education Day, and celebrates the military's guarding of territorial borders for the present and future wellbeing of the nation.
10 This is a government-promoted economic reform model supposed to deliver Thailand from a middle income trap, perceived as overly reliant on low-tech agricultures and low-skilled industry, to advanced economy status based on "creativity and innovation" inspired by "smart people" and an "inclusive society" (http://thailand.prd.go.th/mobile_detail.php?cid=1&nid=3785. Accessed 4 March 2017).

References

Adams, W. M. (1991) Large Scale Irrigation in Northern Nigeria: Performance and Ideology. *Transactions of the Institute of British Geographers*, 16, 287–300.

Allouche, J. (2005) *Water Nationalism: An Explanation of the Past and Present Conflicts in Central Asia, the Middle East and Indian Sub-continents*. Geneva, Institute of International Studies.

Anderson, B. (2006) *Imagined Communities: Reflections on the Origin and Spread of Nationalism*. London, Verso.

Barker, R. and Molle, F. (2004) Evolution of Irrigation in South and Southeast Asia. *Comprehensive Assessment Research Report*. Colombo, Comprehensive Assessment of Water Management in Agriculture.

Berkoff, J. (2007) Some Economic Aspects of Large Rice-based Projects in Southeast Asia. In Facon, T. (Ed.) *The Future of Large Rice-Based Irrigation Systems in Southeast Asia*. Ho Chi Minh City, Viet Nam, FAO Regional Office for Asia and the Pacific.

Blake, D. J. H. (2012) *Irrigationalism: The Politics and Ideology of Irrigation Development in the Nam Songkhram Basin, Northeast Thailand*. PhD thesis, School of International Development, Faculty of Social Sciences. Norwich, University of East Anglia.

Blake, D. J. H. (2015) King Bhumibol: The Symbolicœ 'Father of Water Resources Management' and Hydraulic Development Discourse in Thailand. *Asian Studies Review*, 39, 649–668.

Blake, D. J. H. (2016) Iron Triangles, Rectangles or Golden Pentagons? Understanding Power Relations in Irrigation Development Paradigms of Northeast Thailand and Northern Cambodia. In Blake, D. J. H. and Robins, L. (Eds.) *Dynamics of Water Governance in the Mekong Region*. Petaling Jaya, Malaysia, Strategic Information and Research Development Centre (SIRD).

Blake, D. J. H. (2017) Water Flows Uphill to Power: Hydraulic Development Discourse in Thailand and Power Relations Surrounding Kingship and State Making. In Baghel, R., Stepan, L. and Hill, J. K. W. (Eds.) *Water, Knowledge and Environment in Asia: Epistemologies, Practices and Locales*. Abingdon, Oxon, Routledge.

Blake, D. J. H. and Robins, L. (2016) Introduction: A Backdrop to Water Resources Governance Dynamics in the Mekong Region. In Blake, D. J. H. and Robins, L. (Eds.) *Water Governance Dynamics in the Mekong Region*. Petaling Jaya, Malaysia, Strategic Information and Research Development Centre (SIRD).

Brummelhuis, H. T. (2005) *King of the Waters. Homan van der Heide and the Origin of Modern Irrigation in Siam*. Leiden, KITLV Press.

Callahan, W. A. (2006) History, Identity and Security: Producing and Consuming Nationalism in China. *Critical Asian Studies*, 38, 179–208.

Chaloemtiarana, T. (2007) *Thailand: The Politics of Despotic Paternalism*. Chiang Mai, Silkworm Books.

Chanida, C. (2007) *Khrong-gan an nueang ma jak phraratchadamri: gan-sathapana phraratcha amnat nam [The Royally-Initiated Projects: The Making of Royal Hegemony (B.E. 2494–2546)]*. Faculty of Social Science and Anthropology. Bangkok, Thammasat University.

Chitkasem, M. (1999) Politics and Thai literature. In Mallari-Hall, L. J. and Roxas-Tope, L. R. (Eds.) *Tests and Contexts: Interactions Between Literature and Culture in South East Asia*. Manila, University of the Philippines.

Chomchai, P. (1994) *The United States, The Mekong Committee and Thailand: A Study of American Multilateral and Bilateral Assistance to Northeast Thailand Since the 1950s*. Asian Studies Monographs. Bangkok, Institute of Asian Studies, Chulalongkorn University.

Deutsch, E. and Bontekoe, R. (Eds.) (1997) *A Companion to World Philosophies*. Malden, MA, Wiley-Blackwell.

Fan, S., Jitsuchon, S. and Methakunnuvut, N. (2004) *The Importance of Public Investment for Reducing Rural Poverty in Middle-Income Countries: The Case of Thailand*. Washington, DC, International Food Policy Research Institute (IFPRI).

Faures, J.-M. and Mukherji, A. (2009) *Trends and Drivers of Asian Irrigation*. Colombo, International Water Management Institute (IWMI) and the Food and Agriculture Organisation of the United Nations (FAO).

Floch, P. and Blake, D. (2011) Water Transfer Planning in Northeast Thailand: Rhetoric and Practice. In Lazarus, K., Badenoch, N., Dao, N. and Resurrecion, B. P. (Eds.) *Water Rights and Social Justice in the Mekong Region*. London, Earthscan.

Floch, P. and Molle, F. (2013) Irrigated Agriculture and Rural Change in Northeast Thailand: Reflections on Present Developments. In Daniel, R., Lebel, L. and Manorom, K. (Eds.) *Governing the Mekong: Engaging in the Politics of Knowledge*. Petaling Jaya, Strategic Information and Research Development Centre (SIRD).

Floch, P., Molle, F. and Loiskandl, W. (2007) Marshalling Water Resources: A Chronology of Irrigation Development in the Chi-Mun River Basin, Northeast Thailand. *M-POWER Working Papers*. Chiang Mai, Mekong Program on Water, Environment and Resilience.

Fong, J. (2009). Sacred Nationalism: The Thai Monarchy and Primordial Nation Construction. *Journal of Contemporary Asia*, 39(4), 673–696.

Gladney, D. C. (2004) *Dislocating China: Minorities, Muslims, and Other Subaltern Subjects*. Chicago, The University of Chicago Press.

Grachangnetara, S. and Bumrungtham, P. (2003) Productivity of Rice and Irrigation in Thailand. *TDRI Quarterly Review*, 18, 17–23.

Guyot, E. (1987) *Isan Khieo – The Green Northeast*. Hanover, NH, The Institute of Current World Affairs.

Hall, B., Minami, I., Jantakad, P. and Dinh, C. N. (2014) *Irrigation for Food Security, Poverty Alleviation and Rural Development in the LMB*. Vientiane, Lao PDR, Mekong River Commission.

Hamilton-McKenzie, J. (2009) California Dreaming: Selling the Irrigationist Dream. *Journal of Historical and European Studies*, 2, 27–38.

Hirsch, P. (2014) Laos Mutes Opposition to Controversial Mekong Dam. In Hilton, I. (Ed.) *Forging a New Course for the Mekong*. London, China Dialogue.

Hoanh, C. T., Facon, T., Thuon, T., Bastakoti, R. C., Molle, F. and Phengphasy, F. (2009) Irrigation in the Lower Mekong Basin Countries: The Beginning of a New Era? In Molle, F. F. T. and Käkönen, M. (Eds.) *Contested Waterscapes in the Mekong Region: Hydropower, Livelihoods and Governance*. London, Earthscan.

Htet Naing Zaw. (2016) No New Irrigation Dams During Govt's Term. *The Irrawaddy*. Yangon, Irrawaddy Publishing Group.

204   *David J.H. Blake*

Htoo, Thant. (2016) Govt to Survey Dams Nationwide. *The Myanmar Times*. Yangon, Myanmar Consolidated Media, Ltd.

Intravisit, A. (2005) *The Rhetoric of King Bhumibol's Sufficiency Economy: Rhetorical Analyses of Genre and Burke's Dramaticism of December 4th Speeches of 1997, 1998, 1999 and 2000 Jan–Jun 2005 ed*. Bangkok, Bangkok University.

Isager, L. and Ivarsson, S. (2010) Strengthening the Moral Fibre of the Nation: The King's Sufficiency Economy as Etho-politics. In Ivarsson, S. and Isager, L. (Eds.) *Saying the Unsayable: Monarchy and Democracy in Thailand*. Copenhagen, NIAS Press.

Jackson, P. A. (2010) Virtual Divinity: A 21st-Century Discourse of Thai Royal Influence. In Ivarsson, S. and Isager, L. (Eds.) *Saying the Unsayable: Monarchy and Democracy in Thailand*. Copenhagen, NIAS Press.

Johnston, R. M., Hoanh, C. T., Lacombe, G., Noble, A., Smakhtin, A., Suhardiman, D., Kam, S. P. and Choo, P. S. (2010) *Rethinking Agriculture in the Greater Mekong Subregion; How to Sustainably Meet Food Needs, Enhance Ecosystem Services and Cope with Climate Change*. Colombo, Sri Lanka, International Water Management Institute.

Kaida, Y. (1978) Irrigation and Drainage: Present and Future. In Ishii, Y. (Ed.) *Thailand: A Rice-Growing Society*. Honolulu, The University Press of Hawaii.

Kartin, A. (2001) Water Scarcity Problems in Israel. *GeoJournal*, 273–282.

Keyes, C. F. (1977) *The Golden Peninsula: Culture and Adaptation in Mainland Southeast Asia*. New York, Macmillan Publishing Co., Inc.

Lipchin, C. (2007) Water, Agriculture and Zionism: Exploring the Interface Between Policy and Ideology. In Lipchin, C., Pallant, E., Saranga, D. and Amster, A. (Eds.) *Integrated Water Resources Management and Security in the Middle East*. Netherlands, Springer.

Matthews, N. (2012) Water Grabbing in the Mekong Basin – An Analysis of the Winners and Losers of Thailand's Hydropower Development in Lao PDR. *Water Alternatives*, 5, 392–411.

Matthews, N. and Geheb, K. (Eds.) (2015) *Hydropower Development in the Mekong Region: Political, Socio-economic and Environmental Perspectives*. Abingdon, Oxon, Routledge.

Menga, F. (2015) Building a Nation Through a Dam: The Case of Rogun in Tajikistan. *Nationalities Papers*, 43, 479–494.

Menga, F. (2016) Domestic and International Dimensions of Transboundary Water Politics. *Water Alternatives*, 9, 704–723.

Mirumachi, N. (2012) Domestic Water Policy Implications on International Transboundary Water Development: A Case Study of Thailand. In Ojendal, J., Hansson, S. and Hellberg, S. (Eds.) *Politics and Development in a Transboundary Watershed: The Case of the Lower Mekong Basin*. Dordrecht, Springer.

Molle, F. (2003) Allocating and Accessing Water Resources: Practice and Ideology in the Chao Phraya River Basin. In Molle, F. and Srijantr, T. (Eds.) *Thailand's Rice Bowl: Perspectives on Agricultural and Social Change in the Chao Phraya Delta*. Bangkok, White Lotus.

Molle, F. (2005) Irrigation and Water Policies in the Mekong Region: Current Discourses and Practices. *IWMI Research Report*. Colombo, Sri Lanka, International Water Management Institute (IWMI).

Molle, F. (2007) Scales and Power in River Basin Management: The Chao Phraya River in Thailand. *The Geographical Journal*, 173, 358–373.

Molle, F. (2008) Why Enough Is Never Enough: The Societal Determinants of River Basin 'Overbuilding'. *International Journal of Water Resources Development*, 24, 217–226.

Molle, F. and Floch, P. (2008) Megaprojects and Social and Environmental Changes: The Case of the Thai 'Water Grid'. *Ambio*, 37, 199–204.

Molle, F., Floch, P., Promphakping, B. and Blake, D. J. H. (2009a) The 'Greening of Isaan': Politics, Ideology, and Irrigation Development in the Northeast of Thailand. In Molle, F., Foran T.

and Käkönen, M. (Eds.) *Contested Waterscapes in the Mekong Region: Hydropower, Livelihoods and Governance*. London, Earthscan.

Molle, F., Mollinga, P. P. and Wester, P. (2009b) Hydraulic Bureaucracies and the Hydraulic Mission: Flows of Water, Flows of Power. *Water Alternatives*, 2, 328–349.

Mollinga, P. P. (2008) Water, Politics and Development: Framing a Political Sociology of Water Resources Management. *Water Alternatives*, 1, 7–23.

Mukherji, A., Facon, T., Burke, J., De Fraiture, C., Faures, J.-M., Fuleki, B., Giordano, M., Molden, D. and Shah, T. (2009) *Revitalizing Asia's Irrigation: To Sustainably Meet Tomorrow's Food Needs*. Colombo, Sri Lanka, International Water Management Institute and Food and Agriculture Organisation of the United Nations.

Neef, A. (2008) Lost in Translation: The Participatory Imperative and Local Water Governance in North Thailand and Southwest Germany. *Water Alternatives*, 1, 89–110.

Office of the Prime Minister. (1979) *Thailand Into the 80's, Bangkok, Office of the Prime Minister*. Royal Thai Government.

Ounvichit, T. (2005) People's Participation in Irrigation Management in Thailand. In Shivakoti, G. P., Vermillion, D. L., Lam, W.-F., Ostrom, E., Pradhan, U. and Yoder, R. (Eds.) *Asian Irrigation in Transition*. New Delhi, Thousand Oaks and London, Sage Publications.

People's Irrigation Act B.E. 2482 (1939) http://thailaws.com/law/t_laws/tlaw0195.pdf

Pramoj, M. R. S. (1990) Stone Inscription of Father King Ramkhamhaeng: First Constitution of Thailand. In Ratanakul, P. and Kya Than, U. (Eds.) *Development, Modernization, and Tradition in Southeast Asia: Lessons From Thailand*. Bangkok, Mahidol University.

Reynolds, C. J. (Ed.) (2002) *National Identity and Its Defenders: Thailand Today*. Chiang Mai, Silkworm Books.

Rigg, J. (2005) Poverty and Livelihoods After Full-time Farming: A South-East Asian View. *Asia Pacific Viewpoint*, 46, 173–184.

Rigg, J. and Ritchie, M. (2002) Production, Consumption and Imagination in Rural Thailand. *Journal of Rural Studies*, 18, 359–371.

Samudavanija, C. (2002) State-Identity Creation, State-Building and Civil Society, 1939–1989. In Reynolds, C. (Ed.) *National Identity and Its Defenders: Thailand Today*. Revised ed. Chiang Mai, Silkworm Books.

Sattayanurak, S. (2005) The Construction of Mainstream Thought on 'Thainess' and the 'Truth' Constructed by 'Thainess'. *Conference on Thai Humanities II*. Chiang Mai, Thailand.

Schluter, S. (2006) Motivation and Participation in Irrigation Management in Thailand. In Gaese, H. (Ed.) *Congress on Integrated Water Resources Management*. Berlin, ITT.

Scott, J. C. (1998) *Seeing Like a State: How Certain Schemes to Improve the Human Condition Have Failed*. New Haven and London, Yale University Press.

Sellman, J. D. (2002) *Timing and Rulership in Master Lu's Spring and Summer Annals*. Albany, State University of New York Press.

Sneddon, C. (2015) *Concrete Revolution: Large Dams, Cold War Geopolitics, and the US Bureau of Reclamation*. Chicago, University of Chicago Press.

Streckfuss, D. (2011) *Truth on Trial in Thailand: Defamation, Treason, and Lese-majeste*. London and New York, Routledge.

Surarerks, V. (2006) Muang Fai Communities in Northern Thailand: People's Experiences and Wisdom in Irrigation Management. *Journal of Developments in Sustainable Agriculture*, 1, 44–52.

Walker, A. (2010) Royal Sufficiency and Elite Misrepresentation of Rural Livelihoods. In Ivarsson, S. and Isager, L. (Eds.) *Saying the Unsayable: Monarchy and Democracy in Thailand*. Copenhagen, NIAS Press.

Wangkiat, P. (2016) Damned If You Do, Damned If You Don't. *Bangkok Post*. Bangkok, Post Publishing Group Ltd.

206 *David J.H. Blake*

Whaley, F. (2005) *Water Management in Thailand – Learning From History*. Manila, Asian Development Bank.

Wiener, A. (1972) *The Role of Water in Development: An Analysis of Principles of Comprehensive Planning*. New York, McGraw-Hill.

Wipatayotin, A. and Wangkiat, P. (2016) Mekong Pumps to Ease Isan Drought. *Bangkok Post*. Bangkok, The Post Publishing PCL.

Wittfogel, K. A. (1981) *Oriental Despotism: A Comparative Study of Total Power*. New Haven and London, Vintage Books.

13 Building a dam for China in the Three Gorges region, 1919–1971

Covell F. Meyskens

In the early twentieth century, Chinese leaders began to contemplate building a large dam on the Yangzi River that would generate electricity, boost river transport, and end its history of floods. They paid particular attention to the Three Gorges region, where the river ended its descent from the Sichuan Basin into Hubei province's Jianghan Plain. This chapter examines Chinese efforts to construct a dam near the Three Gorges between 1919 and 1971. It concentrates on this block of time, because the first dam proposal appeared in 1919, and the Gezhouba Dam became that place in 1971. Past scholarship has overlooked this period and focused on the building of the Three Gorges Dam (TGD) in the 1990s (Dai, 1998).

I take a different tack and trace the Chinese government's aspiration to make a TGD to a new understanding of technology's power precipitated by Western imperialist pressures in the mid-nineteenth century. Prior to Western imperialism's arrival, elites in China did not take the rapid expansion of technological power as a primary economic goal (Wong, 1997; Elvin, 1973). This changed when Western imperialists and their scion – Meiji Japan – used technological strength to assert influence over East Asia. In response, elites in China embraced the Western notion that technological self-strengthening was central to a country's economy and that its territory was a repository of resources useful for industrialization (Halsey, 2015; Wu, 2015; Pietz, 2014).

After laying this historical groundwork, the chapter highlights three defining features of Chinese efforts to erect the TGD. First, Chinese elites manifested a belief in the national benefits of technologically redesigning the environment of the entire Yangzi region. Second, Chinese leaders acted like nationalist elites in other late-developing countries and advocated using state power to advance industry (Gerschenkron, 1962: 5–30). Third, Chinese elites repeatedly sought out foreign assistance to offset domestic shortages of industrial capital. These three trends engendered two approaches to technology, or what Thomas Hughes (1983) calls technological styles, which reflected different understandings of how to use experts, labor, and capital.

The first approach is apparent from 1919, when Sun Yat-sen – the founder of China's first modern mass party, the Guomindang (GMD) – initially suggested a dam in the Three Gorges region. For Sun, the dam was a technocratic affair in which technical experts knew best how to make the Yangzi River contribute its aqueous power to strengthening the national economy. It is not surprising that the GMD followed Sun's technocratic approach, when it worked alone on preparatory work for the dam in the 1930s and with the United States in the 1940s. Previous studies have well established that the GMD shared the global penchant of many mid-twentieth-century regimes for technological solutions to national problems (Kirby, 2000; Scott, 1999).

It is more noteworthy that this technocratic trend persisted after the Chinese Communist Party (CCP) defeated the GMD in the Chinese Civil War in 1949 and allied the People's Republic of China (PRC) with the U.S.S.R. against the U.S. in the Cold War, since much scholarship has argued that the CCP under Mao (1949–1976) put greater stress on political goals than technical expertise (Lee, 1990; Goldman, 1981). This was not entirely the case with the TGD. Similar to some other high-priority projects, Chinese and Soviet experts had significant influence over decision-making (Lewis and Li, 1991). Experts' authority waned during the Great Leap Forward (1958–1962) and the late 1960s, when Mao advocated an alternate technological style that was rooted in the Maoist idea that will power and mass mobilization were more powerful contributors to national development than technical proficiency and advanced technology (Riskin, 1987).

Western imperialism and new visions of power

In the Opium War (1839–1842), the British Empire deployed ironclad steamships against China's last dynasty, the Qing. The British routed Qing troops and through unequal treaties obtained authority over parts of Chinese political and economic affairs. In subsequent years, Western imperial powers repeatedly used their technological superiority to exact concessions from China, including the right for British warships to patrol the Yangzi. Starting in the 1860s, regional officials, such as Zeng Guofan and Li Hongzhang, pushed to increase industry in the Self-Strengthening Movement to counter Western pressures. There has been much debate about whether the government's industrialization attempts were a robust rejoinder to the Western challenge or an inept retort of a sclerotic regime (Elman, 2004; Feuerwerker, 1958). Either way, state-led efforts at industrial expansion marked an important shift in Chinese economic statecraft.

For millennia, Chinese elites viewed agriculture as the economy's foundation. The political salience of this age-old principle eroded in the face of imperialist threats, which gradually made the Chinese intelligentsia think that if a country did not bolster its technological power, it risked an end to its history as a sovereign state. As elites gave industry more economic prominence, they came to see that the economy's energy base could no longer primarily consist of recently deceased plant matter *and* human and animal labor (Von Glahn, 2016: 348–399). China had to uncover other materials within its territory to power industrial growth.

Like elsewhere in the world, Chinese industrial boosters in the nineteenth century took coal to be the main new form of energy (Kander and Malamina, 2014: 131–248). To expand the domestic supply, regional officials fostered coal mining and geologists mapped out coal deposits (Wu, 2015; Wright, 2009: 15–114). After hydroelectric power's invention in the 1880s, some Chinese considered dams as another way to generate energy. One of hydroelectricity's earliest proponents was Sun Yat-sen. In 1894, he wrote a letter to Li Hongzhang, a leading light of the Self-Strengthening Movement, in order to bring hydropower to his attention.

Sun's letter does not appear to have had any policy consequences. It however raised ideas about dams that would remain dominant themes in Chinese dam building. First, the government should promote hydropower development. Second, dams were economically advantageous, because they facilitated river transport and provided limitless energy for economic activities. Third, hydropower plants were a valuable supplement to coal, since they were a sustainable source of power and electricity was transmittable to carbon-poor places (Yang, 2005: 19–20).

Imagining an industrial China

In 1911, the Qing dynasty collapsed and China broke up into competing warlord regimes. Amidst this fraught atmosphere, Sun formulated schemes to make real his dream of an industrial China. He made his most grandiose proposal in 1919 in *The International Development of China* (Sun, 1922). In the book, Sun acted like other nationalist visionaries in late-industrializing countries: he laid out a plan to quickly endow China with a broad industrial base (Kemp, 1989). Sun knew that China did not possess the funds, machinery, or technical knowledge to independently industrialize. So, he called on Westerners to offer their technical expertise and presented investment opportunities.

Sun focused on the creation of nationwide industrial systems, such as transportation and communication infrastructure, that reached from China's far west to blue-water ports on the Pacific. Foreign investors could also bankroll new mining, manufacturing, and energy-producing facilities countrywide. In so doing, China would gain an industrial apparatus whose products spanned the gamut from raw materials to complex machinery (Sun, 1922). One project Sun mentioned was the Three Gorges Dam. In his view, the dam's main economic benefits were its immense output of electricity and stimulation of river traffic (Sun, 1922: 73–74). In 1919, Sun also penned an editorial in which he explained that the Three Gorges was an ideal dam location, because it was narrow and 60 feet deep, and it had high stable walls capable of holding a large reservoir. The Three Gorges region had the added advantage of being located where the Yangzi concluded its 500-meter plunge from Sichuan to Hubei.

Like in *The International Development of China*, Sun envisioned the dam as part of a regional development program. The dam would be one of seven Yangzi River hydroelectric stations, which would account for a sixth of the world's hydropower. Anticipating a critique of the cost, Sun cited a foreign expert's estimates of how much the dam would save in the long run, since hydropower plants required less labor than coal mines and their associated railways (Guowuyuan sanxia gongcheng jianshe weiyuanhui, 2005: 3–4). At the time, no foreigners stepped forward to build the TGD. One likely reason is that the country was split into warlord regimes whose competition induced a violent political environment and troubled economic climate (Sheridan, 1966). In addition, there was a shortage of demand for electricity in the region, and so if a foreign investor had paid for the TGD's high construction costs, they could not have counted on much of a return on their investment in the short or mid term. On the other hand, Sun's approach to the TGD foreshadowed later Chinese officials, who envisaged it as a motor for regional economic growth.

A preliminary survey

In 1927, the GMD reestablished a central government in Nanjing. Its hold on power was shaky, as warlords continued to exert influence and Japan persistently sought to take Chinese territory (Eastman, 1974). The GMD nonetheless attempted to build up the national economy using technocratic methods (Strauss, 1998; Kirby, 1984). To oversee development efforts, the GMD founded the National Defense and Design Commission (NDDC) in 1928. In the same year, the GMD set up the Yangzi River Water Management Commission. The GMD staffed these organizations with educated elites whose pedigrees were similar to their institutional kin in other late-developing states. They had studied in advanced industrial countries such as the U.S. or Japan, or they had attended Chinese schools infused with Western techno-science (Seth, 2007; Yeh, 2000).

Many of these individuals shared Sun's aspiration to strengthen China by technologically revamping the Yangzi's economic geography. The NDDC made its first salvo in this direction in 1932 amidst trying circumstances. The year before, 142,000 died in Yangzi floods and the Japanese seized Manchuria. Then, in 1932, Japan attacked Shanghai, and the central government temporarily moved the capital inland to Luoyang in Henan province. The government's finances were also in disarray, and so the NDDC's first attempt to survey the Three Gorges was cut short. Soon after, the NDDC decided the TGD was too important to halt, and so it channeled scarce funds to a survey group (Zhonggong zhongyang dangshi yanjiu shi, 2007: 4; Wanli Changjiang zengkan, 1996: 255–256).

Over the next two years, the group engaged in extensive fieldwork (Guowuyuan sanxia, 2005: 4–8). One participant was Song Xishang, who attended MIT and later received an engineering master's from Brown University (Zhongguo xueshu qikan, 2017). Song summarized the group's findings in a Yangzi Commission report that evinces his cosmopolitan background. The report opens with a passage that depicts China as part of a worldwide network of hydropower experts engaged in constructing dams for the sake of nation building. Song also translates the TGD into the dominant language of global techno-science and places English and Chinese terms next to each other when discussing its main structures.

Song further evinces his support of foreign models of hydrological development when he references the regional activities of the Tennessee Valley Authority (TVA). Song embraced the technocratic mindset of the TVA and overlooked how the TGD would transform the ecology of the Yangzi River (Sneddon, 2015; Klingensmith, 2007; Reisner, 1986; Worster, 1985). He instead called on hydropower experts to make a unified plan that considered the economic benefits that the dam could bring to the Chinese people by improving regional transport, irrigation, and energy supplies (Guowuyuan sanxia, 2005: 4–8).

In another report, Song and his co-authors suggested making the TGD part of a regional industrial complex made up of a large chemical plant and three other hydropower stations further up the Yangzi. Yet, they admitted that the dam was presently inadvisable, since China lacked the necessary resources. In addition, it did not make economic sense to allocate limited resources to the dam, since the economy did not need so much electricity (Wanli Changjiang zengkan, 1996: 255–258).

Partnering with the United States

After the early 1930s, the Chinese state's work on the Three Gorges project stagnated for a decade. In the meantime, another 142,000 perished in the 1935 Yangzi floods (Zhonggong zhongyang dangshi yanjiu shi, 2007: 4). During World War II, Japan occupied the nearby city of Yichang and devised plans for a TGD. Unfortunately, Japanese documentation is not available in translation, and so I will not discuss their activities further. As for the GMD, it fled to Sichuan, where the institutional descendant of the NDDC, the National Resources Commission (NRC), endeavored to industrialize the Southwest (Bian, 2005).

In 1942, the NRC sent experts to the U.S. to study science and technology in preparation for postwar reconstruction. The NRC assigned Berkeley-trained civil engineer Zhang Guangdou to the TVA, where Zhang heard that his former U.S. Bureau of Reclamation colleague, John L. Savage, was going to India to assist with the Bhakra Dam. Zhang suggested the NRC sponsor a follow-up visit to China, and the NRC agreed (Yan, 2016).

Once in Sichuan, Savage learned about American advisor G.R. Paschal's recommendation that China and the U.S. cooperate on the Three Gorges Dam and a large fertilizer factory, whose output could repay U.S. technical assistance. Elated at this idea, Savage went to survey the Three Gorges, accompanied by NRC members and the commander of Yangzi defenses, since Japan was known to launch attacks nearby (Yang, 2016: 26). When Savage reached the Three Gorges, a Chinese source reports he declared: "Your honourable nation's founding father Dr. Sun Yat-sen said it well, the Yangzi really is amazing. This place truly contains a huge amount of hydroelectric resources." Savage camped in the region for ten days amidst the threat of Japanese air raids and feverishly collected survey data, taking as his guide Chinese and Japanese military maps as well as

U.S. Army aerial photos (Yang, 2016: 26–28; Wanli Changjiang zengkan, 1996: 99). Savage then returned to Sichuan and wrote the first comprehensive plan for the integrated use of the Three Gorge's water resources.

Since China was at war, Savage recommended placing dam equipment underground. Savage's plan otherwise stressed the economic gains that would come from the world's largest hydroelectric plant. According to his estimates, the dam would generate 68.4 billion kWh annually and a gross income of $143 million. It would stop Yangzi floods, bring water to over 10 million people, and irrigate over 10 million acres. This agricultural stimulus would contribute $37 million to China's GDP every year. The dam would additionally lift annual freight traffic by 33.2 billion ton miles, reduce freight costs by $192 million, and raise tourism earnings by $5 million. To supervise the project, Savage advised the NRC team up with TVA technocrats and the U.S. Bureau of Reclamation (Wanli Changjiang zengkan, 1996: 114–118, 120–121).

After finishing the plan, Savage returned to the U.S. and expressed his enthusiasm for the TGD to Zhang Guangdao. Zhang thought the project was a nonstarter and voiced his disapproval to the NRC. China's economy had no need for so much electricity. The country lacked the funds and technology to realize the project on its own, and the U.S. would probably not issue China a loan, since Congress hesitated to fund the much less expensive TVA. Even were the U.S. to back the dam, Zhang was still against it, because it would give the U.S. too much control over China's economy.

GMD leader Chiang Kai-shek disagreed, and he had the NRC hammer out a contract with the United States in 1945. According to its terms, the U.S. would loan China $3 billion and a group of American engineers, led by Savage, would supervise the Three Gorges project (Wanli Changjiang zengkan, 1996: 260). Despite his reservations, Zhang returned to China to take part. The TGD was a prime opportunity to advance his lifetime goal of building dams to empower the Chinese nation against imperialism (Yan, 2016: 16).

In accordance with the U.S.-China contract, the NRC posted experts to the U.S. Bureau of Reclamation's headquarters in Denver to elaborate a regional development plan. Chinese participants included recent graduates in the U.S. and members of the NRC, the Yangzi River Commission, and central ministries. In parallel, the NRC founded a project headquarters in Nanjing. Following Savage's advice, the NRC appointed American engineer John Cotton as its director and hired other experts from the TVA, the Coulee Dam, and the U.S. military. The NRC pooled domestic engineering talent as well (Wanli Changjiang zengkan, 1996: 262–263). Savage's initial surveys were inadequate, and so lead planners requested more surveys. Lead planners were however in Nanjing and Denver, and so they had to rely on geological teams in the Three Gorges region to gather information. As a result, planners sometimes had to wait for data before they could research key technical questions and craft dam designs (Wanli Changjiang zengkan, 1996: 267–275).

In 1947, the NRC suddenly terminated its partnership with the United States. The cause was the eruption of the Chinese Civil War between the GMD and the CCP in 1945. World War II had weakened GMD finances, and the civil war drove them further into the red. By 1947, the government lacked dollars to pay for U.S. goods and services, and so the NRC informed the Bureau of Reclamation that China was suspending the Three Gorges project. A subsequent report talked about reviving the dam when funds became available (Wanli Changjiang zengkan, 1996: 275–281). The report was prescient. Once China's finances were in order, the Three Gorges project started back up, but only after the the CCP pushed the GMD off the mainland to Taiwan and allied with the Soviets against the Americans in 1949.

Planning with the Soviets

Upon taking power, the CCP established the Yangzi River Water Conservancy Commission and selected Lin Yishan as its director. Lin's credentials were not based on his technical expertise. Rather like many top CCP leaders, they were based on his Party activities since the early 1930s (Lee, 1990: 47–74). It might thus seem that Lin's appointment is evidence that the CCP's approach to the management of the Yangzi River was not technocratic. While it is true that Lin did not start out as a technocrat, he became one. Like many CCP revolutionaries turned bureaucrats, he took special classes in science and technology relevant to his position in the 1950s. What is more, he stressed technical expertise in his handling of the Three Gorges project (Lin, 2004). It is also worth mentioning that as with other highly valued projects, the CCP retained some GMD-era technocrats in the Yangzi Commission (Zhonggong zhongyang dangshi yanjiu shi, 2007: 6; Feigenbaum, 2003).

In its early years, the Commission focused on flood control. Its first major endeavor was the Yangzi River Flood Diversion Project, which mobilized 300,000 people for flood control work in 1952. In 1953, Mao Zedong toured the Yangzi with Lin Yishan and naval personnel on the Yangzi River Warship, an event Mao feted as China regaining sovereignty over its waterways from imperialism (Fenghuang wang, 2007). During the tour, Mao asked Lin whether the flood problem was resolved. When Lin replied negatively, Mao inquired, "What would solve the scourge of Yangzi River floods?" Lin told Mao about a plan for small reservoirs along the Yangzi. Mao then solicited Lin's opinion about whether the TGD would be better. Lin responded, "We really hope to build the Three Gorges Dam, but we dare not think about it now," because China did not have the wherewithal to execute such a mammoth undertaking. Mao however considered the dam to be the best solution to Yangzi floods, and so he told Lin to conduct a study. Lin then ordered the Yangzi Commission's Engineering Department to investigate whether damming the Min, Wu, Jinsha, and Jialing Rivers in the Southwest would settle the flood problem better than the TGD (Zhonggong zhongyang dangshi yanjiu shi, 2007: 7–8).

In 1954, floods killed 33,000 and damaged 4.27 million homes. After the flood, Mao inspected the situation in South China. While in Hubei, he asked Lin Yishan whether China had the technology to construct the TGD. Lin replied "If the Central Government demands at an earlier time to build [the Three Gorges Dam], it can be done relying on our own technical forces and the help of Soviet experts." Otherwise, construction would have to wait until after China had collaborated with the Soviets on the Danjiangkou Dam on the Han River. Then, China would have the requisite experience to tackle the Three Gorges alone, and so Mao instructed Zhou Enlai to ask the Soviet Premier Nikolai Bulganin for experts to aid with designs for the Yangzi River Basin and the Three Gorges Dam.

Mao's request received a positive response, and in 1955, a group of 143 people comprised of ministerial leaders and technical experts surveyed the Three Gorges for two months. In the end, Soviet and Chinese experts came to different conclusions. The Soviets prioritized energy development and suggested following Lin Yishan's earlier proposal to dam the Min, Jinsha, Jialing, and Wu Rivers. The Chinese, on the other hand, took flood control as the most urgent concern and recommended the Three Gorges Dam. In December 1955, Lin Yishan and the head of the Soviet group presented their proposals to Zhou Enlai. Zhou backed China's, because the Soviet plan would inundate lots of arable land in Sichuan, and because it did not address the flooding of areas downstream from the Three Gorges (Zhonggong zhongyang dangshi yanjiu shi, 2007: 9).

Mao was thrilled to hear about the progress of feasibility studies about the dam. In 1956, he again went to Hubei and swam across the Yangzi River, which inspired him to write the poem "Swimming" in which he mused that one day "walls of stone will stand upstream . . . to hold back . . . the clouds and rain till a smooth lake rises in the narrow gorges," a feat which would cause the legendary goddess of the Gorges "to marvel at a world so changed" (Mao, 1956). By March 1956, over 1,000 surveyors were gathering data for the TGD and other Yangzi projects. By September 1957, geological teams had finished collecting information from over 17,448 hydrological observation stations about the Yangzi's sediment content, historic height, and the volume of water flows (Wanli Changjiang zengkan, 1996: 28).

In May 1956, Lin Yishan published an article in support for the Three Gorges project. His endorsement was much more reserved than Mao's paean to the power of technology and the human will to domesticate the Yangzi and impress the divine. Lin, on the one hand, said that the TGD was the optimal way to end the flood problem, and that China had the technical ability to build it. Lin also echoed earlier assessments of the dam's positive contributions to river transport and regional agricultural and industrial development. On the other hand, Lin noted that although based on Soviet reports, the dam was technically possible, a whole slew of issues were under-investigated, such as silting trends and the best construction machinery and hydropower equipment. Only after solving such fundamental problems could China ensure that the Three Gorges Dam would be cost-efficient, based on solid research, and maximally beneficial to national development. Lin additionally repeated Zhang Guangdou's concern that the dam's electrical output would be much more than the Chinese economy could presently consume (Wanli Changjiang zengkan, 1996: 28–32).

Building a dam for China 215

Fellow hydroelectric administrator Li Rui famously came out against Lin Yishan's argument in favor of the dam. Li disagreed with Lin Yishan's claim that China was technologically ready to build the dam, and he did not think that China had enough money. Yet Li was not diametrically opposed to Lin's viewpoint. Like Lin, Li Rui concluded that the TGD would generate more electricity than the local economy required. He admitted that cables could dispatch excess energy to industry far away. But, the same goal was achievable with small dams closer to industrial areas, and those dams would not lead to the forced relocation of an estimated 2.15 million from the dam's flood zone. Li also asserted that the government should not impose its plans on local communities. It had to make a concerted effort to consult with locals and arrive at a consensus (Wanli Changjiang zengkan, 1996: 35–44).

In January 1958, Mao had Lin and Li present their standpoints at a Central Party meeting in Nanning in Guangxi province. Lin mobilized flood data since the Han Dynasty to argue that the Three Gorges was the correct place to definitively deal with the flood problem. Mao asked if the construction costs could be reduced, and Lin replied not with current technologies. Zhou Enlai interjected it was possible if the dam's hydroelectric output was decreased from 25 million kW to 5 million kW. Lin agreed, but added that it would be a waste to obtain so little energy from such a big dam. Li Rui took an opposing viewpoint. He contended that China would not demand so much electricity for a decade. He also questioned whether the dam would actually prevent floods downstream and warned that China's enemies would surely take such a large dam as a target.

Mao concurred that precautions had to be taken to protect the dam, but he still ordered meeting attendees to "actively prepare" a "completely reliable" regional development plan for the Yangzi River Basin centered on the Three Gorges Dam. As Mao often did, he made Zhou Enlai the supervisor of preparations, and Zhou organized a group of 100-plus experts and Party leaders to inspect select potential dam locations. Upon completion, Zhou declared that with current technology China could build the dam within 15 to 20 years (Zhonggong zhongyang dangshi yanjiu shi, 2007: 10).

Zhou then headed to a meeting in Chengdu, where Central Party and provincial leaders conducted early discussions about the Great Leap Forward campaign to rapidly expand China's industrial base. At the meeting's end, a report advocated building the Three Gorges Dam, since China had the technological capability to do so, and the dam would advance the country's long-term economic goals. But, before beginning construction, it was necessary to conduct preparatory work on all major issues. The report projected completing an initial plan for the Yangzi in 1959 and for the dam sometime between 1959 and 1963 (Zhonggong zhongyang dangshi yanjiu shi, 2007: 37–40).

After the Chengdu meeting, Mao rode a Yangzi steamer with Lin Yishan from Chongqing to the Three Gorges region. While on the boat, Mao emphasized the dam's significance and repeated the common tripartite claim that it would end regional floods, stimulate regional transportation, and increase available electricity. Showing just how important Mao thought the project was, he called Lin the

216 *Covell F. Meyskens*

"king of the Yangzi" and offered to resign from his current post and become his assistant. At the end of their journey, Mao again swam across the Yangzi River. That date, March 30th, would in the late 1960s become the code name for the first dam on the Yangzi River – Gezhouba, or Project 330 (Yang, 2016: 45–47).

A near leap

In the late 1950s, the Great Leap Forward enveloped China in a collective fervor to accelerate economic development according to Maoist methods. According to Mao, the Soviet approach to development was too slow and granted too much authority to bureaucrats and technical experts, and so Mao elaborated a different style of socialist industrialization that would putatively enable China to be more self-reliant and make fuller use of its resource strengths (Mao, 1977). As Mao was wont to do, he gave his strategy for nationwide economic engineering a poetic-sounding name – "walking on two legs" – thereby making a top-down governmental vision, which figured China as one gigantic body, appear in name to be a down-to-earth human-sized approach to development.

According to Mao, economic actors should adjust the amount of labor and capital used in the industrialization process based on their availability. In the case of China, its sizable population could compensate for a capital-poor economy and be channeled towards constructing economic infrastructure (Mao, 1977). Mao embedded this approach to development in his military thought and called for the total mobilization of the populace into a fighting force that, like the Party in the 1930s and 1940s , overcame its enemies not through technical expertise and advanced weaponry, but through sheer mass willpower (Riskin, 1987: 81–256). TGD critic Li Rui denounced the Great Leap's developmental strategy and was imprisoned for two decades (Li, 2016: 308–319).

When it came to the TGD, the Party's turn towards a Maoist developmental path had consequences similar to elsewhere in China. It led to a nationalist emphasis on liberating the masses from oppressive foreign models that stressed high technology and bookish knowledge. The Great Leap's influence is visible in meetings about the dam held by bureaucrats and technical experts in June and October of 1958. During these meetings, attendees surveyed possible dam locations and they discussed technical issues, such as the dam's optimum location and shape as well as the best techniques to ameliorate river transport, generate electricity, control floods, and divert water for irrigation.

But, at the same meetings, they also debated the ideal way to draw on the masses' power, so that they could realize monumental feats of physical strength and create technological inventions of intellectual ingenuity that went beyond what was currently thought possible worldwide. In this vein, attendees exercised their techno-scientific imagination and foresaw a giant mechanical crane that could move mountains and the cultivation of a new generation of workers that arrived at the forefront of techno-science, not through formal schooling and research, but by making the TGD (Wanli Changjiang zengkan, 1996: 49–58).

During the Great Leap, Maoist flights of scientific fantasy fueled other dam projects. The most infamous case is the Yellow River's Sanmenxia Dam, where CCP leaders implemented Mao's voluntarist approach to development and made up for shortages of industrial capital with the mass mobilization of militias. Their efforts erected a dam, but it was structurally unsound, quickly silted up, and eventually could not generate electricity (Pietz, 2014: 211–219, 225–229, 301–302). Less well known is the Danjiangkou Dam in northwest Hubei, which Lin Yishan had suggested as a training ground for the TGD. The Danjiangkou Dam did not end up being the model of technical proficiency that Lin had hoped, as it suffered from similar problems of poor engineering as the Sanmenxia Dam (Hubei sheng wenshi ziliao, 1992: 1–7).

What then of the Three Gorges Dam? The Party Center moved towards building it in September 1959, when plans for the dam and Yangzi Basin were completed, and the head Soviet expert in China told Zhou Enlai that they could prepare to begin construction. But, it was not until April 1960 that the Central Party proposed investing 400 million RMB in the Three Gorges project and suggested starting construction in 1961 (Zhonggong zhongyang dangshi yanjiu shi, 2007: 11). That never happened, as it was already clear to the Party Center that Mao's predicted Great Leap into a prosperous industrial future had actually resulted in a full-on plunge into famine and mass death (Wemheuer, 2014). And so, CCP leaders put the TGD on hold again. By late 1961, most Yangzi Commission personnel were involved in farm work, probably to ward off starvation (Zhonggong zhongyang dangshi yanjiu shi, 2007: 11).

A botched attempt

The Three Gorges project remained dead in the water until 1969, when Hubei provincial officials breathed bureaucratic life back into it. The stimulus was anticipated electricity shortages caused by the CCP's Third Front campaign to ramp up military industry in Central China in the wake of rising Sino-Soviet tensions along the northern border (Naughton, 1988). At the time, Hubei was led by members of the People's Liberation Army, which Mao had deployed in 1968 to repress the Cultural Revolution's Red Guard movement and bring order to the country (Walder and Su, 2003). Both Hubei's First Party Secretary, Zeng Siyu, and the second-in-command, Zhang Tixue, had first taken part in revolutionary activities in their teens in the Party's mountain base areas and subsequently risen up its military arm (Zeng, 2014; Hubei sheng Zhonggong dangshi, 1989).

In March 1969, Zhang invited Yangzi Commission members to survey the Three Gorges. Afterwards, he put forward the idea of building the dam to the vice heads of the Ministry of Water Conservancy and Electric Power Qian Zhengying and Wang Yingxian. Qian and Wang endorsed Zhang's proposal, much to his delight. On a visit to Hubei in May, Mao met with Zhang Tixue and Zeng Siyu and discussed the dam. Mao poured cold water on Zhang's enthusiasm. With Sino-Soviet tensions intensifying, Mao thought "right now is

218 *Covell F. Meyskens*

a time to prepare for war, it is inadvisable to consider this . . . If a basin of water fell on your head, would you be scared?" Not deterred, Zhang looked for other options. A member of the Yangzi Commission suggested the Gezhouba Dam. Like the Three Gorges Dam, it would churn out mammoth amounts of energy, but it was a low dam, and so the consequences of a bombardment would be less disastrous than the destruction of the Three Gorges high dam (Zhonggong zhongyang dangshi yanjiu shi, 2007: 12).

Zhou Enlai arranged to discuss the Three Gorges project at a national planning meeting in February 1970. Cultural Revolution struggles had landed Lin Yishan in an informal prison in Wuhan. Wanting to know his opinion, Zhou had Hubei military leaders secure his release. At the meeting, Lin learned to his great surprise that some officials were pushing to build the Three Gorges Dam while Mao was still alive (Hubeisheng weiyuanhui, 1993: 32).

In October 1970, Zhang and other Hubei leaders reported their ideas to the State Council. During the meeting, Zhang made a military pledge to Zhou Enlai and other Party leaders that "If there was a problem with the Gezhouba Dam, then take my head and hang it from Tiananmen" (Zhonggong zhongyang dangshi yanjiu shi, 2007: 12). Despite their apparent ardor for Gezhouba, Zhang Tixue still preferred the Three Gorges Dam, and so he and Zeng Siyu visited Mao in December. Mao started out by chastising Zhang for hurting so many with the Danjiangkou Dam. He tempered his criticism and said that on the other hand, Zhang did have experience with dams, and Zeng was not bad at fighting battles. Yet, Mao still wondered whether they were ready for the Yangzi's first dam. Zhang responded with a famous Mao quote in which he praised learning by doing. Mao complimented them for having guts.

Zhang then talked about how a decade of preparations had already occurred, and how the dam would realize Mao's poem "Swimming." Mao was not swayed and rejected their proposal, because they did not have adequate funds, scientific information, designs, or construction plans. Also, if an atomic bomb hit the Three Gorges Dam, it could cause flooding all the way to Shanghai. At that point, Zhang began pressing his case for the Gezhouba Dam, stating that he had brought experts to Gezhouba to research it and they were ready to move ahead. Mao gave the okay and enjoined them to talk with Zhou Enlai and cooperate with relevant ministries (Hubeisheng weiyuanhui, 1993: 10–17).

Meanwhile, Zhou Enlai, possibly at Mao's behest, asked Lin Yishan to write a position paper on the Three Gorges and Gezhouba Dam proposals. In November 1970, the State Council held a meeting on the different options. Lin argued that the Three Gorges Dam should be built first, because Gezhouba would cause the water level to rise twenty meters and make the Three Gorges Dam more difficult and expensive (Zhonggong zhongyang dangshi yanjiu shi, 2007: 12). There were also not adequate development plans for Gezhouba. The Party's intention had always been to build the Three Gorges Dam, and so engineers had made designs. Only since 1969 had there been much effort to work on Gezhouba, and that had yet to yield formal designs. In spite of Lin's criticisms, the State Council provisionally approved the construction of Gezhouba (Hubeisheng weiyuanhui, 1993: 20, 33–34, 54).

Zhou however thought Lin's position paper was "a powerful opinion," and so he submitted it to Mao along with the Gezhouba plan, his own letter, and a report, which depicted Gezhouba as realizing the ideal Mao immortalized in his poem "Swimming" of a "smooth lake rising in the narrow gorges." In addition, Gezhouba would fulfill electricity demands, simulate river transport, and overcome Yangzi floods. As a low dam, it posed less of a security challenge, and planning was based on a decade of geographic research and dam modeling experiments for the Three Gorges project, so the past twenty years' mistakes in dam building could be avoided and adjustments could be made during construction (Hubeisheng weiyuanhui, 1993: 4).

Mao read the Gezhouba file on his seventy-seventh birthday, December 26, 1970, and despite the fact that there were insufficient designs, he gave it his imprimatur with the caveat that later revisions should be made as appropriate (Hubeisheng weiyuanhui, 1993: 34). This decision set off of a tidal wave of activity in Hubei, as the provincial government in good Maoist fashion mobilized 100,000 militia members and implemented a policy known as the three simultaneities in which survey, design, and construction work were carried out simultaneously. As the dam took shape, technicians expressed concern to Gezhouba's military leadership about a range of issues from the lack of anti-silting plans and qualified construction workers to laborers pouring concrete without adding ice to stop its expansion. Military leaders excoriated them for betraying the Maoist developmental path (Hubeisheng weiyuanhui, 1993: 34, 75–79).

After a few months, Gezhouba rose above the Yangzi River, but its innards were like a honeycomb (Hubeisheng weiyuanhui, 1993: 69). In late 1972, Zhou Enlai held meetings in which he lambasted project heads, yelling that "If the Yangzi River has a problem, it's not one person's problem, it's the entire country's problem." To prevent Gezhouba's collapse, Zhou demanded workers stop construction immediately and not resume building until technical personnel had made plans that guaranteed the dam's stability. In 1974, construction work started up again. But, by that time, CCP leaders had switched to the American side in the Cold War and begun moving China back towards a more technocratic style of development (Fan and Yang: 67–69; Lee, 1990: 163–328).

Conclusion

Three notable features stand out in Chinese efforts to build a Three Gorges Dam between the 1910s and 1970s. First, Chinese leaders exhibited the late developer's penchant for state-led industrialization. Second, Chinese elites conceived of the Three Gorges Dam as the centerpiece of a program to technologically re-engineer the Yangzi to boost national power and overcome China's position of international weakness. Lastly, due to insufficient domestic capital, elites formed partnerships with more technologically advanced countries. These three trends resulted in two technological styles.

220 *Covell F. Meyskens*

The dominant technological style was technocratic and granted technical experts more authority in the decision-making process. This technological style is evident from Sun Yat-sen's solicitation of foreign aid in the 1910s to the GMD's collaboration with American engineers during the Second World War. After the CCP took power in 1949, the Party gave Chinese and Soviet experts significant influence over the Three Gorges project. Only during the Great Leap Forward did a Maoist technological style gain prominence, which put more stock in mass mobilization and national voluntarism than technical expertise and industrial equipment, an idea that led to a bungled attempt the first time the CCP tried to build a dam in the Three Gorges region in 1971.

Yet, even in this instance, CCP leaders were not entirely indifferent to technical concerns, as evidenced in their queries about whether there were adequate survey data and construction plans. Ultimately, Mao did not listen to the technical advice of the bureaucrat he once anointed the "king of the Yangzi," Lin Yishan. Instead, Mao backed a provincial proposal to build Gezhouba and realize his techno-political dream of a giant wall that tamed the Yangzi River within his lifetime. Once Zhou learned the resultant dam was not viable, he quickly shut the project down and charged Lin and other technocrats with designing a Gezhouba plan in which technical issues held priority. Revisiting a well-worn script of modern Chinese statecraft, they looked abroad for foreign assistance and found the United States (Li, 1999).

Bibliography

Bian, Morris. 2005. *The Making of the State Enterprise System in Modern China: The Dynamics of Institutional Change.* Cambridge: Harvard University Press.

Dai, Qing. 1998. *The River Dragon Has Come!: Three Gorges Dam and the Fate of China's Yangtze River and Its People.* London: Routledge.

Eastman, Lloyd E. 1974. *The Abortive Revolution: China Under Nationalist Rule, 1927–1937.* Cambridge: Harvard University Press.

Elman, Benjamin. 2004. Naval Warfare and the Refraction of China's Self-Strengthening Reforms Into Scientific and Technological Failure, 1865–1895. *Modern Asian Studies.* 38, pp. 283–326.

Elvin, Mark. 1973. *The Pattern of the Chinese Past.* Palo Alto: Stanford University Press.

Fan, Changmin and Yang, Shangpin. 2005. Yichang: Shuidian zhi du. Wuhan: Hubei chubanshe.

Feigenbaum, Evan A. 2003. *China's Techno-Warriors: National Security and Strategic Competition from the Nuclear to the Information Age.* Palo Alto: Stanford University Press.

Feuerwerker, Albert. 1958. *China's Early Industrialization: Sheng Hsuan-huai and Mandarin Enterprise.* Cambridge: Harvard University Press.

Gerschenkron, Alexander. 1962. *Economic Backwardness in Historical Perspective.* Cambridge: Harvard University Press.

Goldman, Merle. 1981. *China's Intellectuals: Advise and Dissent.* Cambridge: Harvard University Press.

Guowuyuan sanxia gongcheng jianshe weiyuanhui. 2005. *Bainian daji: Sanxia gongcheng 1919–1992 nian xinwen xuanji.* Wuhan: Changjiang chubanshe.

Halsey, Stephen R. 2015. *Quest for Power: European Imperialism and the Making of Modern Chinese Statecraft.* Cambridge: Harvard University Press.

Hubeisheng weiyuanhui xuexi wenshi ziliao weiyuanhui. 1992. *Hubei sheng wenshi ziliao, di si shi yi.* Wuhan: Hubei sheng chubanshe.

Hubeisheng weiyuanhui xuexi wenshi ziliao weiyuanhui. 1993. *Hubei sheng wenshi ziliao, di si shi er.* Wuhan: Hubei sheng chubanshe.

Hubeisheng Zhonggong dangshi renwu yanjiu huibian. 1989. *Zhang Tixue zhuan.* Wuhan: Hubei jiaoyu chubanshe.

Hughes, Thomas P. 1983. *Networks of Power: Electrification in Western Society, 1880–1930.* Baltimore: Johns Hopkins University Press.

Kander, Astrid and Paolo Malamina. 2014. *Power to the People: Energy in Europe Over the Last Five Centuries.* Trenton: Princeton University Press.

Kemp, Tom. 1989. *Industrialisation in the Non-Western World.* London: Routledge.

Kirby, William. 1984. *Germany and Republican China.* Palo Alto: Stanford University Press.

Kirby, William C. 2000. Engineering China: The Origins of the Chinese Developmental State. In *Becoming Chinese,* edited by Wen-hsin Yeh, 137–160. Berkeley: University of California Press.

Klingensmith, David. 2007. *"One Valley and a Thousand": Dams, Nationalism, and Development.* New Delhi: Oxford University Press.

Lee, Hong Yong. 1990. *From Revolutionary Cadres to Party Technocrats in Socialist China.* Berkeley: University of California Press.

Lewis, John and Li Xuetai. 1991. *China Builds the Bomb.* Palo Alto: Stanford University Press.

Li, Rui. 1999. Dui lishi fuze daodi: huiyi sanxia gongcheng shangma guocheng de shimou. *Modern China Studies,* 3 [Online]. Available from: www.modernchinastudies.org/us/issues/past-issues/66-mcs-1999-issue-3/504-2012-01-01-10-06-23.html [Accessed 23 July 2017].

Li, Rui. 2016. *Li Rui koushu wangshi.* Hong Kong: Dashan wenhua chubanshe.

Lin, Yishan. 2004. *Lin Yishan huiyi lu.* Beijing: Fangzhi chubanshe.

Mao, Tsetung. 1977. *A Critique of Soviet Economics.* Available from: www.marx2mao.com/Mao/CSE58.html [Accessed 4 July 2017].

Mao, Zedong. 1956. Swimming. *Marxists Internet Archive.* Available from: www.marxists.org/reference/archive/mao/selected-works/poems/poems23.htm [Accessed 27 March 2017].

Naughton, Barry. 1988. The Third Front: Defense Industrialization in Chinese Interior. *The China Quarterly.* 115, pp. 351–386.

Pietz, David. 2014. *The Yellow River: The Problem of Water in Modern China.* Cambridge: Harvard University Press.

Reisner, Marc. 1986. *Cadillac Desert: The American West and Its Disappearing Water.* New York: Viking.

Riskin, Carl. 1987. *China's Political Economy: The Quest for Development Since 1949.* Oxford: Oxford University Press.

Scott, James. 1999. *Seeing Like a State: How Certain Schemes to Improve the Human Condition Have Failed.* New Haven: Yale University Press.

Seth, Sanjay. 2007. *Subject Lessons: The Western Education of Colonial India.* Durham: Duke University Press.

Sheridan, James. 1966. *The Chinese Warlord: The Career of Feng Yu-hsiang.* Palo Alto: Stanford University Press.

Sneddon, Christopher. 2015. *Concrete Revolution: Large Dams, Cold War Geopolitics, and the US Bureau of Reclamation.* Chicago: University of Chicago.

Strauss, Julia C. 1998. *Strong Institutions in Weak Polities: State Building in Republican China, 1927–1940.* Oxford: Oxford University Press.

Sun, Yat-sen. 1922. *The International Development of China.* New York: G. P. Putnam and Sons.

Von Glahn, Richard. 2016. *The Economic History of China: From Antiquity to the Nineteenth Century*. Cambridge: Cambridge University Press.

Walder, Andrew G. and Yang Su. 2003. The Cultural Revolution in the Countryside: Scope, Timing, and Human Impact. *China Quarterly*. 173, pp. 82–107.

Wang, Zhousheng. 2007. 1953 nian Mao Zedong: Fandui diguo zhuyi qinlue yao jianli qiangda haijun. *Fenghuang wang*, 16 July [Online]. Available from: http://news.ifeng.com/history/zhongguoxiandaishi/detail_2011_07/16/7735122_0.shtml [Accessed 4 July 2017].

Wanli Changjiang zengkan. 1996. *Sanxia gongcheng shiliao huibian, 1918–1949*. Wuhan: Changjiang wei xuanchuan xinwen zhongxin.

Wemheuer, Felix. 2014. *Famine Politics in Maoist China and the Soviet Union*. New Haven: Yale University Press.

Wong, R. Bin. 1997. *China Transformed: Historical Change and the Limits of European Experience*. Ithaca: Cornell University Press.

Worster, Donald. 1985. *Rivers of Empire: Water, Aridity, and the Growth of the American West*. New York: Pantheon Books.

Wright, Tim. 2009. *Coal Mining in China's Economy and Society, 1895–1937*. Cambridge: Cambridge University Press.

Wu, Shellen. 2015. *Empires of Coal: Fueling China's Entry Into the Modern World Order, 1860–1919*. Palo Alto: Stanford University Press.

Yan, Guangtao. 2016. *Zhang Guangdou yu sanxia gongcheng jianshe yanjiu, 1944–2009*. Master's thesis, Fujian shifan daxue.

Yeh, Wen-Hsin. 2000. *The Alienated Academy: Culture and Politics in Republican China, 1919–1937*. Cambridge: Harvard University Press.

Zeng, Siyu. 2014. *Bainian jianzheng*. Dalian: Dalian chubanshe.

Zhongguo xueshu qikan. 2017. *Song Xishang* [Online]. Available from: http://xuewen.cnki.net/R2013091280001286.html [Accessed 4 March 2018].

Zhonggong zhongyang dangshi yanjiu shi. 2007. *Zhongguo gongchandang yu changjiang sanxia gongcheng*. Beijing: Zhonggong dangshi chubanshe.

Index

AGUA desalination scheme 40, 43
anticipation 168–169, 173–174, 176, 180–181
apple orchards 139, 140–141
Arizona 1
Asian Development Bank 196
Aswan High Dam 83–84, 87
Aznar, J.M. 38–39

Bangkok 191, 193
Beles River 69–70
Beles Sugar Development Project 66–78
Borçka Dam 105
borderlands 147, 151, 154, 158
Brahmaputra River 1
Bureau of Reclamation (United States) 21–22, 27

California 1, 19–30
capitalist technologies of imagination 168, 170, 179–180
Carlsbad Desalting Plant 25, 28
Central Asia 151–163
Chao Phraya River 192, 193
Chiang Kaishek 212
China 1, 8, 198, 200, 202
Chu-Talas Commission 148, 151–152, 161–162
citizen-shareholders 168, 170–173, 179–180
Cold War 115, 120, 125–126, 129–130
collective interest 96, 99, 101–105, 108–111
Colorado River 1, 19–30
Colorado River Compact 19, 21, 24
communism 192–193
Compagnie National d'Aménagement de la Région du Bas-Rhône et du Languedoc (BRL) 36, 39
consent 100–102, 104–105, 108–111
contestation 99, 101, 107, 111
Çoruh River 105
Cyprus problem 115, 116, 125, 127, 129, 130

Deriner Dam 105
desalination 19–30
developmentalism 99–101, 104–105, 108–111
developmentalist (logic, point of view) 117, 119–120
developmental state 65, 67–78
Druze 131, 138
Dujiangyan Irrigation Scheme 8

Egypt 82–92
encadrement 65, 67, 73, 78
entrepreneurial state 82, 87–92
Estonia 2
Ethiopia 65–78

Flint water crisis 10
Food and Agricultural Organisation 119
Francis turbine 8

Gezhouba Dam 207, 216, 218–220
Golan Heights 131–144
Gramsci 96, 100, 111
Grand Ethiopian Renaissance Dam 90, 92
Greece 2

hegemony 96, 99–101, 105
Herzl, T. 134
high modernism 195, 197, 199
Hoover Dam 6
hydraulic bureaucracy 195
hydraulic mission 82–90, 131, 133–136, 151, 189, 191, 200
hydro-hegemony 147
hydropower nation 169–170, 173, 176, 179–180
hydrosocial territory 131, 133, 136, 138–141

ideology 192, 195, 199, 200, 201
independentist movement (Catalonia) 38–44

224   *Index*

India 1
Indonesia 9
infrastructure 131, 133; counter-infrastructure 131, 132, 136, 139–144; state infrastructure 133, 136–139, 141
initial public offering (IPO) 167–171, 174, 178, 181
inter-communal conflict (Cyprus) 115, 120, 122, 126
international humanitarian law 137, 143
irrigationalism 194, 195, 200
Israel 131, 134, 137–140, 142–144

Jawlani (Syrian Golani) 131–132, 137–144
Jewish National Fund 136–137
Jordan River 134–135, 138–139

Kazakhstan 147–163
King Bhumibol (Rama IX) 192, 198–199, 201
King Mengrai 197, 201
Kyrgyzstan 147–163

Lake Mead 1
Lake Ram 138–141
land reclamation 83–88
Languedoc 35–37, 42
Las Vegas 19, 28
late industrialization 207, 209–210, 219
Lin Yishan 213–215, 217–219
Lipchin, C. 135
Llobregat River 35–37, 40
Lower Mekong Basin 190

Makarios 120, 125–126, 129
Maoism 208, 216–217, 219–220
Mao Zedong 208, 213–220
Marseille 41
Mas, A. 42
Mekerot 136–143
Mekong River 189, 193
membrane desalination 21
Mexico 1, 2
Mey Golan 137–138, 140
modernisation 83–84, 86–87, 96–97, 99–105, 109–111
modern water 49, 50, 57
Mohamed Ali 83–84, 86–87
multi-stage flash distillation 21, 26
Munzur River 99
Muratlı Dam 105

Napoleon 83
Nasser, G.A. 83–85

Nasserism 81
National Hydrologic Plan (Spain) 38–40, 44
national-popular outlook 100–101, 104
National Resources Commission (China) 212–213
National Water Carrier (Israel) 135, 138
nation-state 3–7
Nepal: climate change in 169–170, 174, 181; earthquake 168–169, 173–174, 182; energy emergency 169, 177–179, 184; landslides 173, 174, 175, 180–181; seismic risk 169–170, 174–176, 179–181
Nevada 1
New Orleans 1
New Water Culture (Spain) 38–39
Nile Basin Initiative 87, 90
Nile River 86, 89–90
Non-aligned Movement 125–126
Northeast Thailand 193, 197

occupied Golan Heights *see* Golan Heights
Occupy movement 6
Ottoman 82–83, 92

Palestine 132–136, 140, 143
peacekeeping (in Cyprus) 122–124
periphery 74, 77
Philippines 1
Pyrenees 37–38, 42
Pyrenees-Mediterranean Eurorregion 42

resettlement 75
resistance 13–14, 97–99, 107–108, 112
resource affect 168, 172, 180, 182
reverse osmosis 21, 24, 27
Rio Grande 23
Rosarito Beach 19, 20, 25, 28–29
Royal Irrigation Department 190, 197, 199
Rutenberg concession 132–134, 136, 140, 143

San Diego 21, 26, 28
San Diego County Water Authority 28–29
Savage, J. 211–212
scale and water 6–10
scarcity 53, 57, 59
settler colonialism 131–137, 142–144
shareholder model 168–170, 170–173, 179, 180–181
social costs 62
social relations 49–51, 55, 58–62
Soviet Union 208, 213–214, 216–217, 220
sugar production 67, 69
Sun Yatsen 207–209, 220

Index   225

surveillance 55, 60
Syria 131, 135–139, 143

technocracy 207–208, 210–213, 219–220
technology 207–208, 210–216, 219–220
territorialisation 66, 70, 78, 132, 135–136, 143; de-territorialisation 131–134, 138; re-territorialisation 133–134
Ter River 35–36, 38, 40, 43
Têt River 49–51, 53–56, 58–59, 61–62
Thainess 196, 201
Thorp, W.L. 116–120, 124–125, 127–128, 130
Three Gorges Dam 207–208, 201–220
Tijuana River 23
Toshka 88
Turkey 96–111

Udall Plan for the Pacific Southwest 25
U.K. 5
uncertainty 51, 55, 58, 62
United Nations 5, 116, 122–123, 129–130

United Nations Economic Commission for Asia and the Far East 193
United Nations Economic Commission for Europe 152–153
United States 1, 208, 211–213

Vinça Dam 49, 50, 57–58, 61

water cooperatives 140–142
Water Framework Directive 43
water nationalism 82–84, 87, 89–92
water rights 140, 143
waterscape 147, 163
World Bank 97, 194
World Commission on Dams 6

Yemen 9
Yuma 26, 27

Zhou Enlai 214–215, 218–220
Zionism 132, 134–136, 143

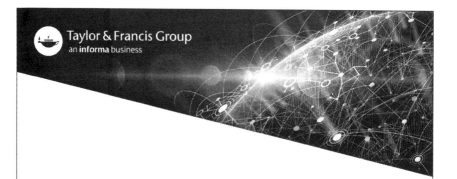